BIOLOGICAL
AND MEDICAL PHYSICS SERIES

Springer
Berlin
Heidelberg
New York
Hong Kong
London
Milan
Paris
Tokyo

Physics and Astronomy ONLINE LIBRARY

http://www.springer.de/phys/

BIOLOGICAL
AND MEDICAL PHYSICS SERIES

The field of biological and medical physics is a broad, multidisciplinary, and dynamic one, touching on many areas of research in physics, biology, chemistry, and medicine. The Biological and Medical Physics Series is intended to be comprehensive, covering a broad range of topics important to the study of biological and medical physics. Its goal is to provide scientists, medical doctors and engineers with text books, monographs and reference books to address the growing need for information.

Wan-Li Xing Jing Cheng (Eds.)

Biochips

Technology and Applications

With 50 Figures

Springer

Dr. Wan-Li Xing
18 Life Science Parkway,
Changping District,
Beijing 102206,
P. R. China
e-mail: wlxing@capitalbiochip.com

Dr. Jing Cheng
18 Life Science Parkway,
Changping District,
Beijing 102206,
P. R. China
e-mail: jcheng@capitalbiochip.com

Cataloging-in-Publication Data applied for

Bibliographic information published by Die Deutsche Bibliothek Die Deutsche Bibliothek lists this publication in the Deutsche Nationalbibliografie; detailed bibliographic data is available in the Internet at <http://dnb.ddb.de>

ISSN 1618-7210

ISBN 3-540-00423-8 Springer-Verlag Berlin Heidelberg New York

Springer-Verlag Berlin Heidelberg New York
a member of BertelsmannSpringer Science+Business Media GmbH

http://www.springer.de

© Springer-Verlag Berlin Heidelberg 2003
Printed in Germany

Data conversion: Marianne Schillinger-Dietrich, Berlin
Cover concept by eStudio Calamar Steinen using a background picture from The Protein Databank (1 Kzu). Courtesy of Dr. Antoine M. van Oijen, Department of Molecular Physics, Huygens Laboratory, Leiden University, The Netherlands. Reprinted with permission from Science 285 (1999) 400–402 ("Unraveling the Electronic Structure of Individual Photosynthetic Pigment-Protein Complexes", by A. M. van Oijen et al.) Copyright 1999, American Association for the Advancement of Science.
Cover production: *design & production* GmbH, Heidelberg

Printed on acid-free paper   57/3141/ts - 5 4 3 2 1 0

Preface

Biochips represent a revolutionary technology that will have significant impact on the science, and indeed life of the human beings in the twenty-first century. This book presents contributions from more than ten internationally renowned experts in the field of biochips. It covers topics such as microarray technology and its applications, microfluidics, drug discovery, detection technology, lab-on-chip technology and bioinformatics. The idea for publishing this book arose at the International Forum on Biochip Technologies 2000 (IFBT 2000) held in Beijing. This is the first international forum dedicated to the new scientific and technological developments in biochip technologies. More than 300 scientists in the area of biochip technologies attended the forum to communicate their latest research results and to discuss the development of biochip technology. After the forum, twelve outstanding experts attending the forum from both academia and industry were invited to prepare manuscripts for this book. Here the authors not only describe their latest research work, but also discuss the current trends in biochip technology. We believe that both researchers in the field of life sciences and professionals in the field of biochips will benefit from this book.

February, 2003

Jing Cheng
National Engineering Research Center
for Beijing Biochip Technology
Beijing, China

Contents

Oligonucleotide Array Technologies for Gene Expression Profiling

David Dorris, Chang-gong Liu, Ramesh Ramakrishnan, Richard Shippy, Sangeet Singh-Gasson, Anna Lublinsky, Edward Touma, Marc Domanus, Luis Allegri, Hong Fei and Abhijit Mazumder

Abstract

We report on expression profiling technologies that employ arrays of oligonucleotides covalently attached to polymeric surfaces. These surfaces can be comprised of photochemically polymerized slabs or film-type structures. We demonstrate that the expression profiling technologies can be used for toxicology and biological discovery. Refinement of the entire platform (probe design, array manufacturing, assays and software analysis) has led to a minimal detectable sensitivity of one copy per cell, a dynamic range of two to three logs, a specificity of greater than 90%, a minimal fold change of 1.8-2 fold and the majority of CVs below 40%. The performance, throughputs and flexibility of this system should enable widespread use in biological and pharmaceutical applications.

1. Introduction

Gene expression analysis on a genome-wide scale can provide insights into cellular physiology after genetic, environmental, chemical or pathological perturbation (Lockhart and Winzeler 2000; Young 2000). Fluorescently labeled cRNA or cDNA hybridized to microarrays of oligonucleotides (Lockhart and Winzeler 2000) or cDNA clones (Schena et al. 1995) are a powerful tool for the analysis of gene expression. Microarrays detect gene expression levels in parallel by measuring the hybridization of cRNA or cDNA to many thousands of genes immobilized on a glass surface (the "chip"), providing a sensitive, high throughput method well suited to take advantage of the sequence and clones produced in genome sequencing efforts.

When gene sequence information is available oligonucleotides can be synthesized to hybridize specifically to each gene in the sample. This approach obviates the need for management of large clone libraries as it is guided primarily by sequence data. Furthermore, use of oligonucleotides is favorable for representation of rarely used splicing patterns (which would be hard to find as cloned cDNAs no matter how carefully the libraries are normalized) and for distinguishing between closely related members of gene families. This approach is particularly suited to analyzing the expression profiles of organisms with completely sequenced genomes as all predicted genes can be analyzed.

Oligonucleotides can be synthesized in-situ or prefabricated and then printed. Synthesis of oligonucleotides by light-directed, combinatorial solid phase chemistry (Fodor et al. 1991) or other in situ methods (Maskos and Southern 1992) offers the advantage of having the oligonucleotide synthesized on the support to be used in the hybridization, obviating the need to hydrolyze the oligonucleotide from its synthetic support and reattach it to the microarray. However, this approach does not allow independent confirmation of the fidelity of synthesis nor does it allow purification of the oligonucleotide prior to attachment to the microarray. Moreover, due to the lower yields of many of these in situ synthetic protocols, oligonucleotides synthesized to date are generally not longer than 25 bases. Furthermore, photomasks are required for photolithographic microarray design.

Covalent attachment of prefabricated oligodeoxyribonucleotides circumvents these restrictions and allows new elements to be added without redesigning the entire microarray. The primary concern with postsynthetic attachment is whether a robust, specific and reproducible attachment chemistry can be created to yield high sensitivity and reproducibility in the subsequent assays. We demonstrate a manufacturable array technology which, when coupled to improvements in analysis tools and assay formats, yields sensitive, reproducible and reliable measurements of transcript levels and changes in those levels.

2. Materials and Methods

CodeLink™ Expression Bioarrays. Target is defined as the labeled nucleic acid sample applied to the CodeLink™ DNA microarray. Probe refers to the DNA oligonucleotide covalently attached to a gel matrix. Each gene is represented by three probes on the microarray. Target was labeled by incorporation of biotinylated nucleotides. The biotin labeled cRNA was purified by Qiagen RNeasy column and cRNA quantitated by measuring the absorbance at 260 nm. Target was then hybridized to the probe microarray. After an 18 hour incubation at 37 °C, arrays were washed for 30 minutes at 42 °C in TNT buffer (150 mM Tris-Cl, pH 7.5, 0.1% Tween-20), then processed using the CodeLink Expression developer kit.

The CodeLink Expression Bioarrays were scanned at 10 µm resolution using the Motorola[R] CodeLink Scanner. The microarray images were analyzed with the CodeLink Analysis software for semiautomatic gridding and automatic quantitation.

Yeast experiments. The standard lab strain S288C was grown in liquid YPD medium containing 2% dextrose at 30°C, then split into two equal cultures. One culture was subjected to a 30 minute heat shock at 37°C. Then RNA was harvested by standard techniques (Ausubel et al. 2000). Poly (A)$^+$ RNA was prepared using the Qiagen Oligotex kit (Qiagen), then used to prepare labeled cRNA.

10,000,000
1,000,000
100,000
10,000
1,000
Fluorescence
3 K
300
30K
100K
300K
cutoff
0.1
1
10
100
1000
Spike Concentration (pM)

A

10,000,000
1,000,000
100,000
10,000
Fluorescence
PM
1 M M
2 M M
3 M M
4 M M
Cutoff
AF061442:1098
AF106860:1134
AF106860:167
ENTF:1163
Perfect Match (PM) and Mismatch (MM) Probes

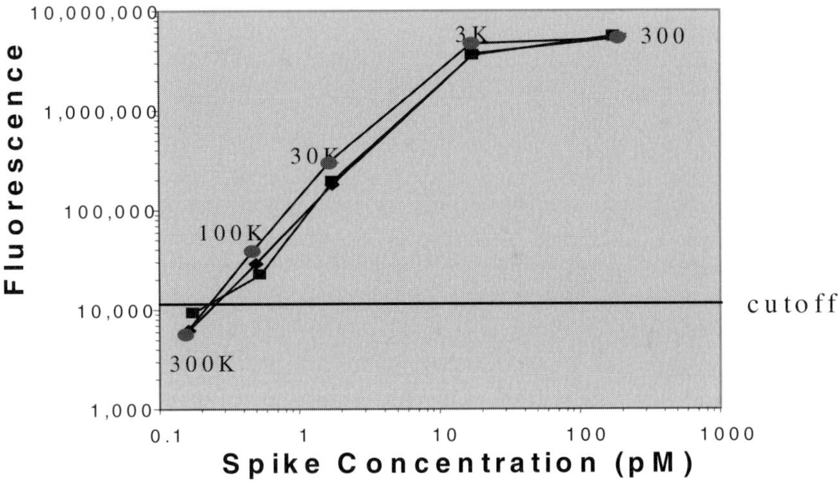

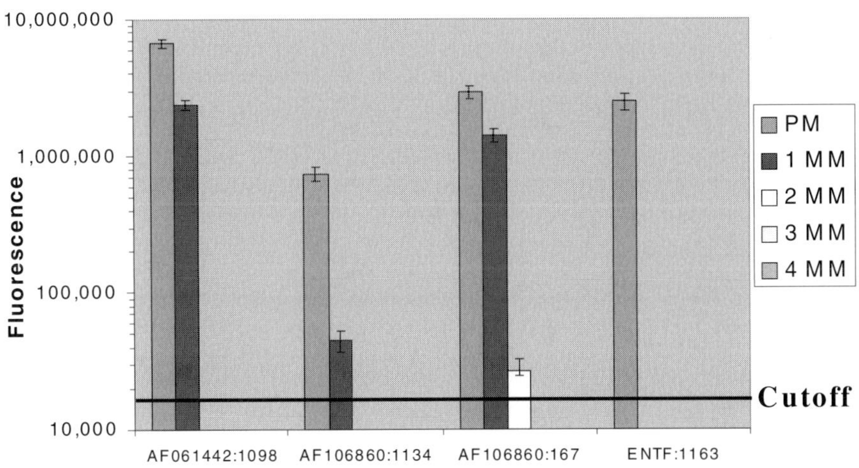

B

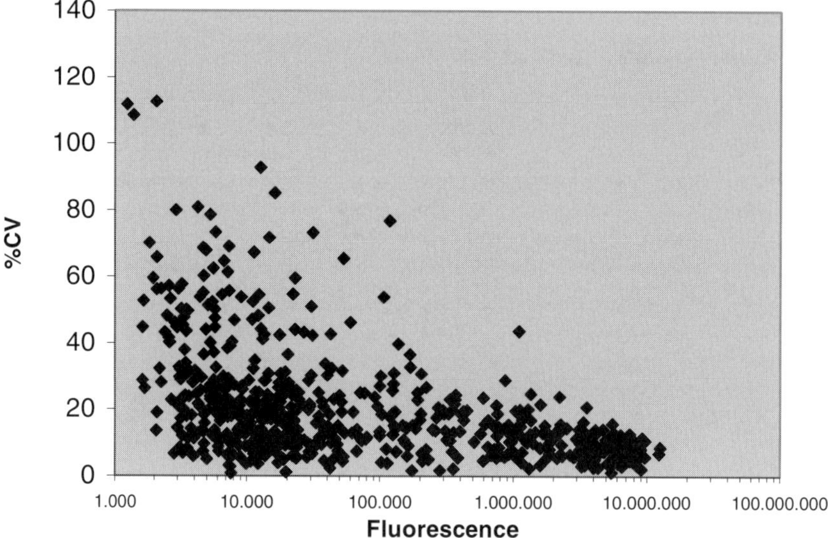

C

Fig. 1. Sensitivity, specificity and CVs from typical design verification experiments performed with the ADME bioarray. (**A**) Sensitivity determined by spiking bacterial transcripts into rat poly A+ RNA at various mass ratios. The hybridization signal for each mass ratio is shown. The squares, diamonds and circles correspond to the entF, fixB and yjeK transcripts, respectively. (**B**) Specificity and ability to distinguish closely homologous genes is demonstrated by analyzing the effects of mismatches on the hybridization signal for various 30 mer probes. In each case, two or more mismatches reduced the hybridization signal for that reaction to 1% or less of that of the wild-type case. (**C**) CV plotted as a function of hybridization signal for a typical experiment. The CV was determined from a set of eight hybridizations.

Photochemical attachment experiments. Probes directed against eight different yeast mRNAs were included on the chip to determine sensitivity. One µg of human placental poly(A)⁺ RNA was spiked with 17 pg of yeast poly(A)⁺ RNA.

3. Results and Discussion

Demonstration of assay performance. The first bioarray product was a low density chip (113 genes; 3 probes per gene) designed to measure the transcript levels of genes involved in drug metabolism. This ADME (absorption, distribution, metabolism and elimination) chip contained phase 1 and 2 metabolizing enzymes, housekeeping genes, transporters and many others. Such applications will allow researchers to gain molecular insights into toxicity profiles for lead compounds (Waring and Ulrich 2000). Excellent sensitivity (one to two copies per cell), specificity (greater than 90%), and CVs (majority lower than 40%) were demonstrated on this chip (Fig. 1). The specificity observed using this technology was especially critical because of the highly homologous nature of many of the genes on the chip (e.g., members of the cytochrome P450, glutathione transferases and others). In contrast, cDNA microarrays typically offer specificity in the 70%-80% range and would not offer the discrimination necessary for analyzing such homologous genes.

In addition to polymeric film-type structures, an earlier chip platform used polyacrylamide slab matrices. Indeed, one of our earlier chip platforms was constructed by photochemically attaching synthetic oligonucleotides on three dimensional surface of a polyacrylamide based hydrogel. We have also demonstrated good sensitivity with this platform. Figure 2 shows a plot of the fluorescence intensity of different yeast transcripts detected when each was spiked into human placental poly (A)+ RNA at a mass ratio of 1 to 300,000 (equivalent to one copy per cell). Seven out of eight probes (each probe corresponding to a different yeast transcript) were significantly over the background cutoff fluorescence signal. Cutoff is defined as the mean signal of the blank pads plus three standard deviations which gives a 99.7% likelihood of having a real signal. The sensitivity of the assay on this platform has thus been shown to be down to 17 pg of starting poly (A)+ RNA. Therefore, the assay and oligonucleotide array formats are compatible with and perform well on both polymeric film-type structures and polyacrylamide slab matrices.

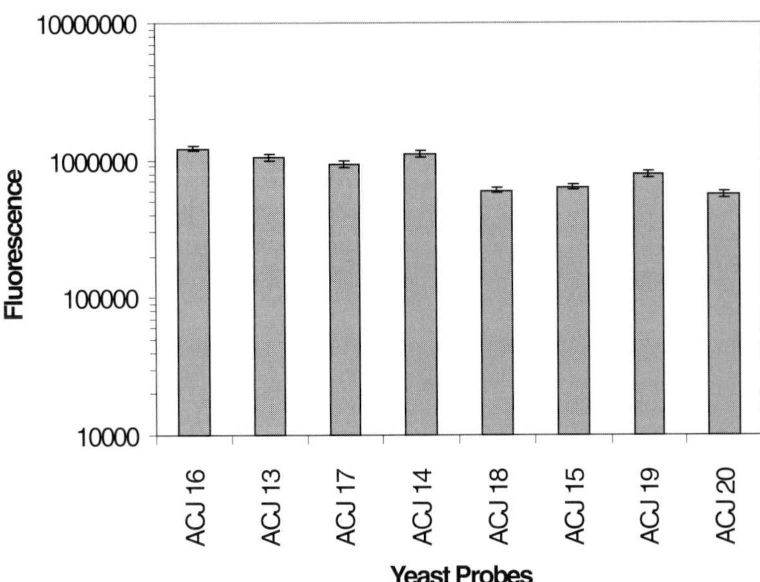

Fig. 2. The same target sample (with yeast mRNA spikes) was hybridized to 8 separate CodeLink Expression Bioarray chips. The mean and standard deviation is shown for each probe hybridization signal. The cutoff was determined to be the average background plus three standard deviations.

Verification and validation of assay results. The budding yeast Saccharomyces cerevisiae serves as a model eukaryotic organism. It is the first eukaryotic organism to have its genome fully sequenced and is also the first organism to have every open reading frame represented on a DNA microarray (DeRisi et al. 1997; Goffeau 1996). Numerous transcriptional profiling studies have been performed on yeast, many of which are available in public databases (Aach et al. 2002; DeRisi et al. 1997). Budding yeasts are also amenable to standard molecular techniques. These facts make budding yeast an excellent model system to validate and study the new CodeLink™ Expression System.

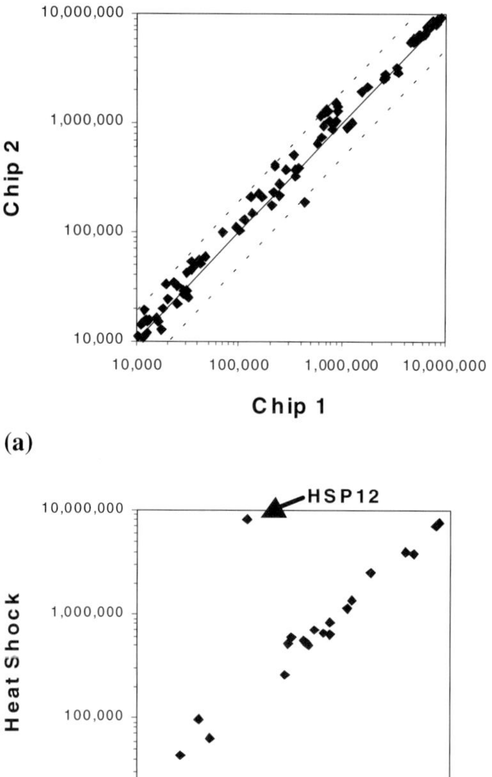

(a)

(b)

Fig. 3. (a) The same target sample was hybridized to 2 separate CodeLink Expression Bioarray chips. 99% of the points are within 2-fold in this chip-to-chip comparison. The dotted line represents a 2-fold difference in signal. **(b)** RNA was prepared from cells grown at 30°C or after a shift to 37°C for 30 minutes. The fluorescent intensity for each gene is represented in both panels.

Probes corresponding to a number of yeast genes were used to construct DNA microarrays. These genes include constitutively expressed "housekeeping" genes and differentially regulated genes. A standard laboratory yeast strain was subjected to normal and heat shock conditions to test the performance of the CodeLink Expression Bioarrays. First, the microarray chip to chip variability was measured by hybridizing the same target sample to multiple microarrays (Fig.3a). This minimal-fold detectable difference experiment for the CodeLink Expression System shows that 99% of the probes are within a 2-fold difference when comparing one microarray chip to another microarray chip (Fig. 3a). Figure 3b shows a plot of transcript levels from cells grown under normal conditions compared to cells grown under heat shock conditions. As expected, the heat shock gene HSP12 is activated after the heat shock. The majority of genes do not change their expression level after the temperature shift.

Once again, the CodeLink Expression Bioarrays were found to perform at very high levels. The sensitivity, selectivity, dynamic range, and minimal-fold detectable difference performance were comparable to competitive systems commercially available (Lockhart and Winzeler 2000). Furthermore, the CodeLink Expression System was used to detect differential transcript levels in a biological sample (Fig. 3b).

Analysis of transcript levels in the budding yeast show that differential gene expression can be monitored using the CodeLink Expression System. In addition to results presented in Fig. 3a, the CodeLink Expression System can detect differential gene expression in yeast cells grown under various growth conditions such as diauxic shift, galactose induction, etc. (data not shown). These results in budding yeast suggest that any biological sample can be assayed using the CodeLink Expression System to measure mRNA transcript levels.

Future optimizations. The standard concentration for robotic dispensing is 18 μM oligonucleotide probe. To further enhance CodeLink expression chip performance and evaluate the maximum capacity of probe attachment on the Gen 3 platform, a series of experiments was performed using different approaches, such as varying the thickness of the Gen 3 polymer or varying the dispensing concentration. Data from variable thicknesses indicated that increasing thickness of the gel increased signal intensities of expressed genes and sensitivity (data not shown). Furthermore, increasing dispensing concentration was found to have effects on signal intensity, sensitivity and specificity. Fig. 4 shows the change of signal intensities on an oligo probe that is complementary to a yeast mRNA target present at mass ratios of 1 in 300, 1 in 3000, 1 in 30,000 and 1 in 300,000. This probe was arrayed on an expression chip where the probe was dispensed in five different concentrations (18, 36, 72, 144 and 288 μM). In all cases, the signal increases 30-50% when the dispensing concentration is increased.

Next, to study methods for improved specificity and discrimination on this low density expression chip a series of mismatch probes with one base, two bases, three bases and four bases at the central positions of 30 mers oligo probes were designed and built on the chip. The results in Fig. 5a indicate that the discrimination improved as the probe concentration increased. This result follows basic nucleic acid hybridization kinetics as shown in Fig. 5b. As the Cot curves depict, better discrimination is obtained with increasing probe concentrations. This scenario is observed on the microarrays tested in this study as well (Fig. 5a).

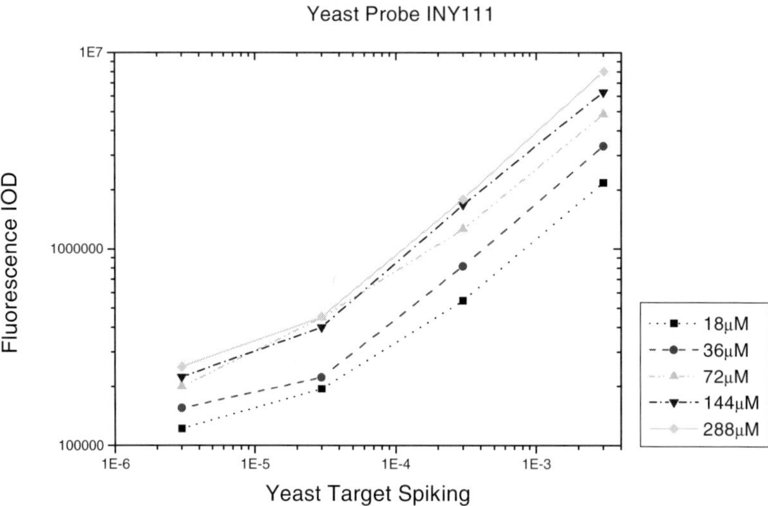

Fig. 4. Signal intensities of a transcript spiked into human mRNA at different mass ratios as a function of spike concentration for five different oligo dispensing concentrations. Probes were dispensed at five different concentrations and hybridized transcripts were detected using biotinylated human cRNA.

Analysis software. Motorola has developed analysis software to extract bio-array data from images and to output gene expression data for gene expression profiling experiments. CodeLink software is one of the components in the family of CodeLink System components that Motorola BioChip Systems has been developing for its microarray platform. Other modules in the CodeLink System Software family include software for scanning, for data mining and for SNP genotyping.

CodeLink-Analysis extracts data from a bioarray by generating a synthetic grid and then fitting it to an actual array using probe-landmarks. After laying down the grid, the software extracts the probe signal and determines noise thresholds. Subsequently, the software generates probe data reports and analysis data in which rules/heuristics determine the exact state of a gene. Comparative analysis can be performed between different arrays/samples through normalized data. The program generates graphical reports for specific gene signal comparisons and graphical reports for sample versus sample comparisons.

CodeLink-Analysis DB is a software package that integrates bioarray data analysis, CodeLink-Analysis, and data storage and retrieval. Powerful queries can be performed on multiple bioarrays. Results are generated based on statistical summaries. One of its uses is in performance monitoring of manufactured bioarrays. By grouping bioarrays into small QC-batches the system can perform statistical data summaries that provide information on the quality of the product.

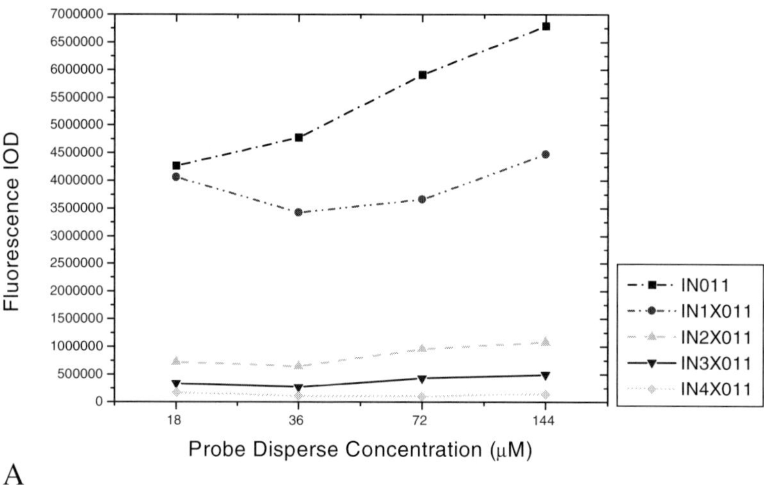

A

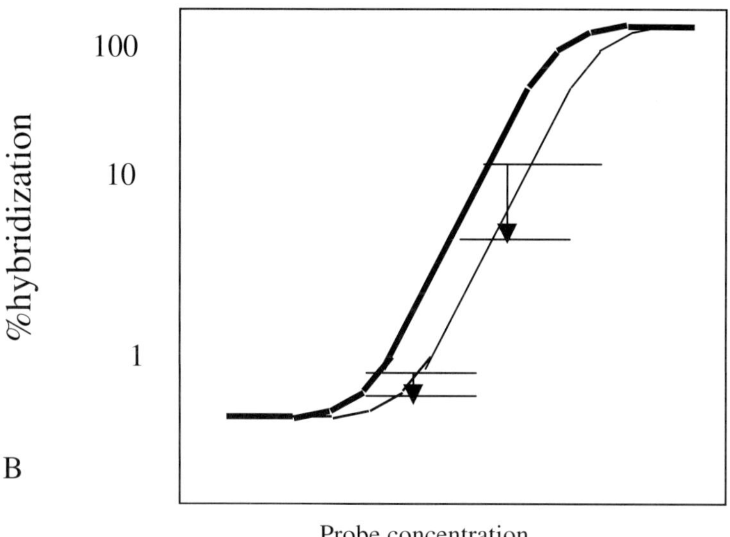

B

Fig. 5. Mismatch discrimination. (**A**) Effect of oligonucleotide dispensing concentration on mismatch discrimination. Perfect match probe 011 and its mismatch probes with one base (1X), two bases (2X), three bases (3X) and four bases (4X) at the central position of a 30 mer oligo were dispensed at four different concentrations as indicated. The discrimination follows basic nucleic acid kinetics as depicted in a Cot curve. (**B**) Schematic diagram of a Cot curve for matched (dark line) and mismatched (thin line) duplexes. The arrow indicates the extent of discrimination and increases as concentration increases.

In summary, we have presented a novel, three-dimensional bioarray platform for sensitive and reliable gene expression profiling measurements. The sensitivity (one copy per cell), dynamic range (2.5 to 3 logs), minimal detectable fold change (99% of data points within 2-fold and 95% within 1.8-fold), low CVs in hybridization signals (majority less than 40%), specificity (90% or greater) and low amounts of target required (50 ng of poly A+ RNA) demonstrate the utility of the polymeric surface and covalent attachment methodologies. Platforms such as these will empower researchers to discover new pharmaceutically relevant targets and elucidate functions and pathways of novel proteins.

References

Aach J, Rindone W and Church GM Systematic management and analysis of yeast gene expression data. Genome Research 10:431-445

Ausubel FM, Brent R, Kingston RE, Moore DD, Seidman JG, Smith JA and Struhl K (eds) (2000) Current protocols in molecular biology. John Wiley & Sons, New York

DeRisi JL, Iyer VR and Brown PO (1997) Exploring the metabolic and genetic control of gene expression on a genomic scale. Science 278: 680-686

Fodor SPA, Read JL, Pirrung MC, Stryer L, Lu AT and Solas D (1991) Light-directed, spatially addressable parallel chemical synthesis. Science 25: 767-773

Goffeau A et al (1996) Life with 6000 genes. Science 274:546

Lockhart DJ, Dong H, Byrne MC, Follettie MT, Gallo MV, Chee MS, Mittmann M, Wang C, Kobayashi M, Horton H and Brown EL (1996) Expression monitoring by hybridization to high-density oligonucleotide arrays. Nature Biotechnology 14: 1675-1680

Lockhart DJ and Winzeler EA (2000) Genomics, gene expression and DNA arrays. Nature 405: 827-836

Maskos U and Southern EM (1992) Oligonucleotide hybridizations on glass supports: a novel linker for oligonucleotide synthesis and hybridization properties of oligonucleotides synthesized in situ. Nuc Acids Res 20:1679-1684.

Schena M, Shalon D, Davis RW and Brown PO (1995) Quantitative monitoring of gene expression patterns with a complementary DNA microarray. Science 270: 467 470

Waring JF and Ulrich RG (2000) The impact of genomics-based technologies on drug safety evaluation. Annual Rev Pharmacol Toxicol 40: 335-352

Young RA (2000) Biomedical discovery with DNA arrays. Cell 102: 9-15

Medical Therapy: The Next Frontier of Biochip and Biomedical Nanotechnology

Ferrari M., Friend D., Hansford D., Kulkarni S. S., Martin F.

Abstract

In the recent past rapid strides have been made to exploit the applications of nanotechnology in biomedical applications. While the focus of such research has primarily been on diagnostic tools, the therapeutic applications have not been so widely researched. This chapter seeks to review the state of the art of nanotechnology in therapeutic applications.

The manipulation of conventional lithographic techniques to produce monodisperse nanoporous, biocompatible, silicon membranes has been described. The potential use of these microfabricated nanoporous membranes in site-specific drug delivery applications has been investigated and their use in the fabrication of immunoisolating biocapsules has been elucidated. Specifically, an insulin-delivering microfabricated biocapsule has been described. Using the same nanotechnlogy principles, an innovative, nanotechnology-based systemic treatment method for metastatic deposits, employing a microfabricated Natural Killer (NK) mimicking cell has been propounded.

1. Introduction

This chapter primarily concerns the possibility of producing a drug delivery system based on nanotechnology that can be manufactured in mass and at an affordable cost. The beginning of the chapter provides an introduction to the bio micro-electro-mechanical systems technology that makes such a system possible. The rest of the paper is dedicated to demonstrating how conventional lithography may be employed to generate – very reliably and reproducibly - critical features for the delivery system with dimensions as small as *ten nanometers*. The basic technology is described in section two and current biomedical applications in the field of drug delivery and implantable, immunoprotected cell bioreactors are reviewed in section three. These sections are prefaced by a brief review of therapeutic bioMEMS and nanotechnology and a discussion identifying some potential areas in which nanotechnology offers significant advantages over existing therapeutic modalities. The chapter concludes with an elaboration on a proposed innovative, nanotechnology-based systemic treatment method for metastatic deposits.

1.1   MEMS and Biological Systems

Massive investments by the Defense Advanced Research Project Agency (DARPA) of the United States fueled the development of Micro-Electro-Mechanical Systems (MEMS) technology in the 1980's. Predicated on the notion of integrating electronic intelligence with mechanical sensing and actuation, the field of MEMS has developed into a bona-fide industry sector of its own, with global market size estimates exceeding the billion US dollar threshold in 1998, and growing (NEXUS! (1996-2002) Marketanalysisfor microsystems http://nexus-emsto.com/ market/ analysis). The further combination of MEMS with biological elements, with intended applications in the life and clinical disciplines, has given rise to the novel, exceedingly promising set of new enabling technological platforms, collectively known by the self-explanatory acronym "bioMEMS". In our taxonomy, bioMEMS include microfabricated biosensors, microfluidic systems and biochips.

Biosensors consist of electronic elements transductively coupled with biomolecules or whole cell biological detecting units. Microfluidic systems are typically fabricated by methods that employ photolithography, either directly for the definition of the channel architectures in silicon, silicon oxide and polymers, or indirectly, in the formation of the silicon or glass templates from which polymer replicas are derived. Also known as micro Total Analysis Systems (microTAS), microfluidic platforms have engendered a major overhaul of the medical diagnostic sector, bringing about the notion of point-of-care diagnostics through miniaturized, portable "laboratories" (Bergveld 2000). A particularly elegant use of photolithography in the life sciences has been the development of light-defined, spatially directed assemblies of multi-biomolecular arrays (McGall et al. 1999) – a technology that has touched off a veritable revolution in the field of sequencing of nucleic acids, with potentially unimaginable upsides in health care in general. BioMEMS for medical *therapeutic* applications are establishing themselves as a powerful driving force for creative innovation. In view of this chapter's specific emphasis on therapeutic micro- and nanosystems, some major accomplishments in this field are reviewed in more detail below.

A feature that is common, in broad strokes, to biosensors, microfluidics, biochips, and bioMEMS in general is the fact that photolithographic processes are employed in their fabrication. These processes by their very nature lend themselves to efficient, economic mass-scale replication, as convincingly demonstrated by the microelectronics industry. On these bases it may then be argued that for the field of bioMEMS, as well, there is a possible pathway towards mass commercial fruition - at least as long as lithographic processes are employed in their manufacture.

Fairly recent advances in photolithography have lowered the spatial resolution limit to below the micron threshold. Devices may thus be routinely produced that comprise features with lateral geometries in the range of hundreds of nanometers – thus true 'nanotechnology' obtained by conventional 'top-down', mass-production scalable methods. To be sure, several extremely exciting new approaches have given access to nanoscale constructs for biomedical applications, including carbon nanotube technology, nanomanipulation by atomic force microscopy, nonstandard lithographies and nanoparticle systems. With the advent of methodologies that will hopefully permit the economically feasible scale-up of such nano-processing methods, it may be confidently expected that they will yield truly revolutionary advances in health care. For the time being, however, a case may be made for the strategic convenience of addressing the nanoscale world in biomedicine through conventional photolithography – with the benefit of a recognizable pathway to commercial scale-up.

1.2 Therapeutic BioMEMS and Nanotechnology

Peripheral nerve information transducers (Edell 1986), cortical implants for the evoking of images in visually impaired patients(Campbell et al. 1991)and micro-electrodes for the stimulation of the central auditory system(Anderson et al. 1989)were among the first therapeutic microdevices proposed. Taking advantage of the truly atomic-level sharpness of cutting edges obtained by crystallographically selective chemical etches of single-crystalline silicon, high-precision micro-surgical instruments were fabricated at Lawrence Livermore National Laboratories (Microfabricated Devices at the Lawrence Livermore National Labs http://www.llnl. gov/eng/eetd/mtc / Devices html #microgripper).

Perhaps the majority of R&D activities in biomedical therapeutic uses of microdevices and nanotechnology, however, focuses on drug delivery. Santini et al. (1999) recently disclosed a microchip that can be triggered by the application of an electrical potential and releases molecules in a tightly controllable and programmable pulsatile fashion. In its current embodiment this microsystem is perhaps still a few generations away from a truly implantable, therapeutic device, but nevertheless it demonstrates features that have been eluding drug delivery researchers for decades: the ability to release single or multiple menus in a truly pulsatile fashion. Fundamental elements of a technology that may generate portable systems for on-field drug reconstitution and delivery were demonstrated by Evans et al. (1997).

Chen and Wise (1994) first proposed micro machined needles that could be employed for the delivery of therapeutic molecules. Microsystems for the *transdermal* delivery of soluble drugs have received substantial interest in recent times, with companies such as ALZA (Macroflux™), Kumar and Redeon developing product lines in the sector. Micromachined needle arrays for transdermal delivery were proposed by Henry (1997), Henry et al. (1998), Pisano and associates (Lin and Pisano 1999; Talbot and Pisano 1998), Mc Allister et al. (1998), and Brazzle at el. (1998).

Micromachined probes on stents, for the potential delivery of anti-restenosis agents into coronary agents were presented by Reed et al (1998). Intravascularly (IV) injectable, active delivery *microparticles* were envisioned and actually patented by Benjamin et al. (Benjamin and White 1988), but were never reduced to practice, possibly in view of the untimely passing away of the lead inventor. Multifunctional, micromachined IV particles were also investigated at Berkeley (Ferrari 2000) and micron-sized, prototypical particles were successfully demonstrated together with the conjugation chemistry necessary for their derivatization with amine groups and immunoglobulin G (Nashat et al. 1998). Current embodiments of injectable multifunctional microparticles are discussed in the section entitled "Intravascular, Systemic Treatment of Metastatic Disease", below.

To our knowledge, the first example of inorganic *nano*technology for the delivery of therapeutic molecules was developed by one of the authors and his group at Berkeley. In this context, the key enabling technology is the fabrication of pores of *absolute* size as small as 10 nanometers. Such dimensions are beyond the *theoretical* limit of resolution of photolithography. The stratagem that was employed to overcome this difficulty involves the use of a sacrificial layer technique (see Fig. 1): A sacrificial layer is grown or deposited onto a substrate material and a subsequent covering layer is deposited on the sacrificial stratum. Prior to the deposition of the covering layer, the sacrificial material is patterned, so as to provide regions of contact and anchorage between the substrate and the covering upon deposition of the latter. The covering is then patterned to etch holes that allow the irrigation of the sacrificial layer with a chemical that etches it, but affects neither the substrate nor the covering layer. Thus, a channel is formed between the substrate and the covering, such substrate having the exact dimensions of the sacrificial layer. In view of the fact that such layers can be deposited with atomic layer control over their thickness, if desired, and with < 5% variability in their thickness over standard silicon substrates, channel dimensions as small as ten nanometers are readily obtained. And the technology required is quite trivial, widely available and reliably reproducible on a large scale, a *top-down nanotechnology*, precisely.

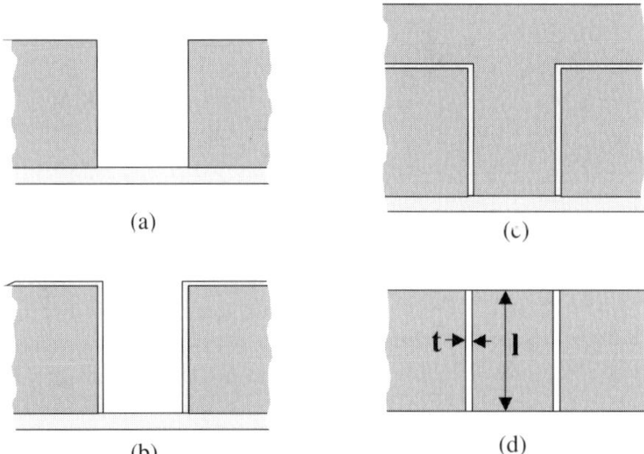

Fig. 1. Summary of fabrication steps: **(a)** etching of holes in structural silicon layer; **(b)** growth of sacrificial oxide by dry oxidation; **(c)** deposition of second structural silicon layer; and **(d)** planarization of second silicon layer and removal of sacrificial oxide and protective nitride layers

1.3 Target Issues in Drug Delivery

The field of drug delivery has attracted considerable attention over the past 30 years beginning with the founding of Alza and Elan. The promise of drug delivery has, in some regards, been fulfilled although clear success of the more advanced applications (e.g., drug targeting) remains elusive. In fact, the technologies used in current commercial drug delivery systems were developed many years ago. For example, the first proposed use of osmotic pump technology for oral and other applications dates back to the late 1960s and early 1970s. Likewise, transdermal dosage forms have been available since the late 1970s. Newer drug delivery technologies currently attracting investment and attention, including inhaled and sustained release injectable peptide/protein drugs from biodegradable polymers, have been under development since the mid to late 1980s.

As we move into the next century it seems reasonable to scan the horizon in search of the next generation of drug delivery systems. At least conceptually, these newer approaches should address those unsolved issues that still engage the drug delivery scientist. These include drug targeting (of peptides, proteins, and DNA) to improve cancer chemotherapy and cardiovascular treatment, improved control of the rate of release from implanted dosage forms, pulsatile drug delivery, closed-loop systems that can sense a change in the body's biochemistry leading to modulated rate or extent of drug delivered and the ability to implant xenogeneic cells in humans to more naturally control a range of diseases, including diabetes and neurodegenerative conditions.

A quick perusal of journals dealing with drug delivery reveal that the older approaches, such as the use of liposomes, nanoparticles, sustained release microspheres and the like are still being pursued to solve the problems outlined above. Yet these approaches have been studied in great detail over the past 20 years with little substantial progress.

Drug delivery, like many applied sciences, draws upon different disciplines to solve its problems. Traditional pharmaceutical approaches have been central to the field; strong contributions from chemical engineering, biomaterials and a range of medical specialties have also been made. The field of bioengineering is emerging as a scientific discipline in its own right. It is from this field, in the form of BioMEMS, that we can expect new and exciting solutions to some current drug delivery problems. This technology offers a wide range of novel attributes that can be incorporated in drug delivery systems. As noted above, microfabrication allows the manufacture of uniform particles and membranes down to nearly a single nanometer. For instance, membranes that can sequester cells away from the body's immune system while allowing the transport of appropriate molecules have been prepared with pores as small as 5 nm with no more than 5% ($\pm$0.25 nm) variability (Ferrari et al. 1996). No other method is currently available to prepare such small and uniform pores. While these membranes have been used for immunoisolation purposes, they can also be used as rate limiting membranes for peptide and protein drugs. Preliminary data (unpublished) from a study with 24.5 nm pore membranes and a 5 μm pore length demonstrated zero-order release of a model peptide (lysozyme) for about 2 weeks under in vitro conditions. Unlike biodegradable polymeric microsphere systems, there was no burst of drug release, obviating the need for elaborate formulation efforts. The peptide (an enzyme) was stable over the course of the experiment. It is possible to prepare membranes in a variety of pore sizes and configurations to accommodate a wide range of molecular weight drugs (from under 500 to 200,000 daltons).

The ability to control particle size is another unique feature of microfabrication technology. It is relatively simple to prepare small (down to a micron or below) particles that are essentially identical to each other. This is not the case with other particulate drug carriers where invariably a range of particle sizes is produced as an inherent attribute of the method of manufacture. In addition to size, defined pores or reservoirs can be introduced into these particles. These reservoirs can hold drug in either a solid or solution/suspension state until the target site is reached.

The problem of drug targeting can be addressed through the use of asymmetric ligand binding and unidirectional drug release. Modifying a single surface of a microfabricated particle through existing chemical methods offers the potential for improved delivery to a target epithelial or endothelial surface. This concept is more completely discussed below in the section on intravascular systemic treatment of metastatic disease.

A similar approach to the oral delivery of macromolecules is being pursued by iMEDD. Oral delivery of macromolecules has been limited due to poor permeability and significant drug metabolism in the lumen and at the brush border. Traditional dosage forms can be used to release a drug and adjuvants into the lumen to enhance stability and/or permeability. However, these agents are quickly diluted and hence are relatively ineffective except at high and potentially unsafe concentrations. A BioMEMS-based delivery system can be prepared with multiple agents (drug and adjuvants) incorporated in the same or different compartments. Likewise, release of the adjuvants can be modulated in a manner to maximize bioavailability by the sequential release of the agents from the particles or device. All this can be accomplished in a focal, unidirectional manner by using known binding mechanisms in the gastrointestinal tract. As a result, release of the agents occurs in close proximity to the absorbing surface (the intestinal epithelium).

As noted above, a microfabricated chip has been disclosed wherein it is possible to release a bioactive agent in a highly controlled, pulsatile manner. It would be advantageous to accomplish the same effect without the use of an electronic power supply and from materials that are biocompatible or even biodegradable. Significant strides are required to merge the field of biomaterials with that of microfabrication technology. Studies are underway to develop micromachining protocols for preparation of polymeric, BioMEMS-based drug delivery systems. In this case, the polymers are known to be biocompatible and in some instances are biodegradable.

2. Nanotechnology via Conventional Lithography

Several research groups have used conventional lithography to directly pattern a filtration membrane on a membrane for microfiltration (e.g. (Yang et al. 1998)). While this allows a simple fabrication process, it also greatly limits the minimum features that can be used for filtration. State of the art photolithography is still limited to 250 nm features, so pores smaller than this cannot be produced using standard photolithography. In fact, most university and government research micro-fabrication facilities use older equipment, which limits them to larger feature sizes for entire wafer processes.

Another approach to reducing the pore size using photolithography is the use of interference lithography to produce microfiltration membranes. By using a collimated laser source and a reflecting mirror at an angle to the substrate, they produced an interference pattern on the photoresist-coated wafer. By using interference patterns at $90°$ to each other, a two-dimensional pattern of 260 nm holes was produced with a spacing of 510 nm. The minimum hole size that could be fabricated using this system (Ar^+ laser focused through a pinhole 1.7 m from the wafer) was 175 nm, still above the pore sizes needed for direct interaction with biomolecules (Van Rijn et al. 1998).

Other research groups have recognized the potential of creative microfabrication for defining pores in membrane structures. A group previously at Chalmers University of Technology (Göteberg, Sweden) used a sacrificial oxide to define a flow channel between two silicon membranes. The fabrication process gave a self-aligned filter based on the etch-stop created by heavily boron-doping an opened silicon substrate. While this process has many of the advantages of a simple fabrication scheme and control over pore sizes, it had problems of doping control, pore density considerations and a tortuous flow path (Kittilsland et al. 1990).

Research at the Biomedical Microdevices Center at the University of California, Berkeley, focused on the use of microfabricated devices with nanopores for size-based separation of biomolecules (Ferrari et al. 1996). While the overall design has gone through several generations, the basic structure and fabrication protocol for the nanopores has remained the same, as shown in Fig. 1-3. Photolithography and reactive ion etching are used to define structures in the first silicon layer, which is then oxidized in a dry oxygen environment. This highly controlled oxidation step is what defines the pore thickness, and thus the maximum dimension of molecule that can pass freely through the membrane. The diffusion length of the pore is controlled by the thickness of the membrane, which also affects the stability of the membrane.

By using a thermally grown silicon oxide sandwiched between two structural layers of silicon (either single crystal or polycrystalline), nanopores can be fabricated in silicon structures by selectively etching the sacrificial silicon oxide in a highly selective etchant (HF) (Chu and Ferrari 1998). As shown in Fig. 1(d), the pore thickness (t) and length (l) are highly controlled in this protocol. The third dimension (width) is also controlled by the mask used for the etch through the first silicon layer. Because the HF chemical etch does not attack the silicon, the remaining pore size is exactly the thickness of the sacrificial oxide (Hansford et al. 2001).

The technology of using a sacrificial oxide to define a pore thickness can be used for producing either nanoporous membranes (Desai et al. 1999a) or bonded wafer high-pressure filters (Tu et al. 1999). Both of these structures have uses in biomolecular handling, for controlled passage of molecules via diffusion (membranes) or for ultrafiltration of proteins to remove viruses (bonded wafer filters).

These filters have demonstrated the use of nanoporous structures for the high-pressure separation of particles at the nanoscale. Nanoporous filters with 40 nm nominal pore size were fabricated with a pore variation of < 4%. For 500 nm and 100 nm beads, there was > 99.99% retention and there was > 99.3% retention of 44 nm beads. The slightly lower retention of the smaller beads is explained by the lack of an absolute bead size from the manufacturer, as the beads were sized within 10% of the nominal pore size (40 nm). These results show that nanoporous filters can be designed which would meet the stringent requirements of biofiltration (Tu et al. 1999).

Fig. 2. Backside view of nanoporous membrane on a bulk silicon support for use as an immunoisolation caspule

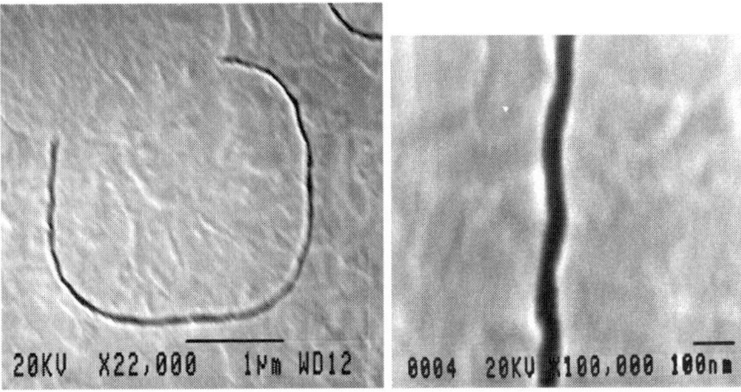

Fig. 3. Micrographs of 50 nm pore showing: (left) single pore from a square pattern mask; and (right) higher magnification of pore showing uniformity of pore thickness even around convoluted shape

3. Applications of Nanopore Technology in Drug Delivery

3.1 Introduction

Recent developments in the field of nanopore technology have led to vigorous research in its biomedical applications. Most of these applications have been directed towards the use of nanopores in diagnostic tools such as labs-on-chips (Fodor 1997), DNA sequencing (Deamer and Akeson), chromatographic, biosensor and cell manipulation systems (Fujimasa 1993; Gourley 1996; McConnell et al. 1992; Volkmuth et al. 1995). However, the vast potential of nanopore technology in therapeutics has not been as widely researched. Some researchers have attempted to use microfabrication techniques in therapeutics and notable efforts include the previously disclosed pulsatile drug delivery microchip (Santini et al. 1999), CNS stimulation (Gu et al. 1999; Wise et al. 1990)and neural regeneration (Akin and Najafi 1994; Kovacs et al. 1994).

A novel application of nanopore technology has been its use in the fabrication of nanopore membranes that can be used in biological environments. These membranes find applications in bioseparations such as blood fractionation, viral filtration and the production of immunoisolated cell bioreactors for the delivery of drugs such as insulin.

3.2 Nanoporous Silicon-Based Membranes as Biofilters

In biological applications it is highly desirable to separate 20-100 nm particles based on their size. An issue that assumes importance in such cases is the fact that membrane filters have to be selective to ratios of 10^4 or higher as opposed to industrial filtrations that require up to 99% filtration. Under these circumstances, a uniform membrane pore-size distribution assumes significant importance. Typically, polymeric membranes have been used in such applications and the pores in these membranes have been formed using either solvent casting or iontrack etching (Lonsdale and Membr 1982).However, it is to be noted here that the pore formation process is stochastic and a size distribution has to be used to describe the actual pore sizes. As a direct consequence, the resulting filtration product will not conform to the biologically required standards of purity. Another limiting factor is that only low densities of tightly controlled pores are achievable, thereby severely limiting the operating flow rates.

Recent research has proved that micromachined silicon can be used to manufacture semi-permeable membranes (Kittilsland et al. 1990; Van Rijn and Elwenspoek 1995; Yang et al. 1998). The mechanical and thermochemical stability of silicon also work in its favor especially in bioseparations involving high pressures. These findings have provided researchers the opportunity to use conventional lithography techniques, as described in an earlier section, to produce membranes that have more desirable properties than the conventionally used polymeric membranes. The authors' group at Berkeley has also performed biologically motivated characterization studies on such a membrane (Desai et al. 1999a). These studies have revealed that the main advantages of using semi-permeable silicon membranes are its biocompatibility (Ferrari et al. 1996) and ease of surface modification (Zhang and Ferrari 1997). Moreover, it has also been seen that the pore-size distribution is uniform and that sterilization is easily achieved. Diffusion studies using beads showed absolute retention and those involving insulin and glucose provided promising results. This last result has led to the use of such a microfabricated nanoporous membrane in the development of an immunoisolation biocapsule that can be used in site-specific drug delivery.

3.3 Immunoisolation and Cell Encapsulation Basics

Immunoisolated cell therapy had first been suggested around thirty years ago in an effort to protect transplanted tissue from immune rejection by enclosing it in a semi-permeable membrane (Lim and Sun 1980). Since then it has been advocated in therapy involving both xenogeneic (across species) and allogeneic (same species) cells encapsulated within a medical device fabricated from known biocompatible materials. The encapsulation membrane serves to allow small molecules such as nutrients to reach the cells and inhibits the passage of larger entities such as antibodies and other immune system constituents. This issue is especially important in xenografts wherein the grafted cells could be very quickly rendered ineffective by the immune system thereby defeating the purpose of the therapy.

Cell encapsulation has since been viewed as holding promise in applications involving site-specific drug continuous delivery of cell-synthesized molecules. Some of the commonly quoted cases in which cell encapsulation therapy has been advocated involve central nervous system disorders such as Parkinson's disease (McGall et al. 1999; Roberts et al. 1996), Alzheimer's disease (Winn et al. 1994) and chronic pain syndrome (Sagen et al. 1993).Other applications involve the performing of replacement organ functions. The most prominent among these has been the treatment of diabetes through islet cell encapsulation and in the treatment of liver dysfunction through hepatic cell encapsulation.

As with any emerging technology, there are still a number of challenges facing cell encapsulation therapy and these are as follows:

- Continued and optimal tissue/cell supply
- Maintenance of cell viability and function
- Successful prevention of immune rejection

Strategies to circumvent these challenges have been offered by Colton (Colton 1996).

Traditionally, as with membrane filters, polymeric materials had been used in the fabrication of the encapsulating membrane.A comprehensive review of these materials has been provided in the form of a review by Li (Li 1998). However, the discovery that nanoporous silicon-based membranes have significant advantages as described earlier has led to the fabrication of a silicon-based microfabricated immunoisolating biocapsule with specific application in the treatment of diabetes.

3.4   Microfabricated Silicon-Based Immunoisolating Biocapsule

Microfabrication techniques that have been routinely used in the electronics industry (MEMS) have been found to provide extraordinary manufacturing precision in the fabrication of silicon membranes for use in biological environments and the same have already been characterized. These very membranes have been used in the fabrication of a biocapsule consisting of two identical micromachined silicon half-capsules. These half-capsules seeded with cells are then bonded together by a thin layer of medical grade silicon elastomer adhesive. The cell containing recess in each half-capsule is made by anisotropically etching into boron-doped (p^+) single crystalline silicon. The recess is then bounded by a surface micromachined polycrystalline silicon nanopore membrane. This membrane serves to provide the size-based biomolecule perm-selectivity essential in immunoisolation. The essential steps in the fabrication of such a capsule are shown in Fig. 4.

The capsule described above has been used in the encapsulation of islet cells for the treatment of diabetes. Rat islet cells have been seeded in each half-capsule and incubated under standard cell-culture conditions. The capsule was then formed by bonding together each half-capsule. Figure 5 shows the schematic of one such fully assembled immunoisolating biocapsule.

Importantly, it has been seen that these anchorage dependent cells grow without any marked changes on silicon surfaces. Moreover, it has been shown that encapsulated islets have shown insulin release profiles better than or at least equal to those of unencapsulated islets (Desai et al. 1998) and these cells have also maintained their stability in vivo for periods similar to, if not longer than those of unencapsulated islets (Desai et al. 1999b). Apart from establishing biofabrication processing and post-processing protocols, and also proving the biocompatibility of such capsules, further studies have also been performed on these capsules relating to the establishment of mass-transport models.

It has thus been proved that the silicon-based immunoisolating biocapsule provides a promising alternative to conventional cell encapsulation techniques with respect to immunoisolation while at the same time maintaining the sufficient diffusion rates necessary for small molecules such as insulin, oxygen and glucose. Thus nanopore technology seems to be poised at a promising stage with respect to intelligent site-specific drug delivery. Further evidence supporting this is provided in the final section probing the creation of a microfabricated artificial Natural Killer (NK) cell.

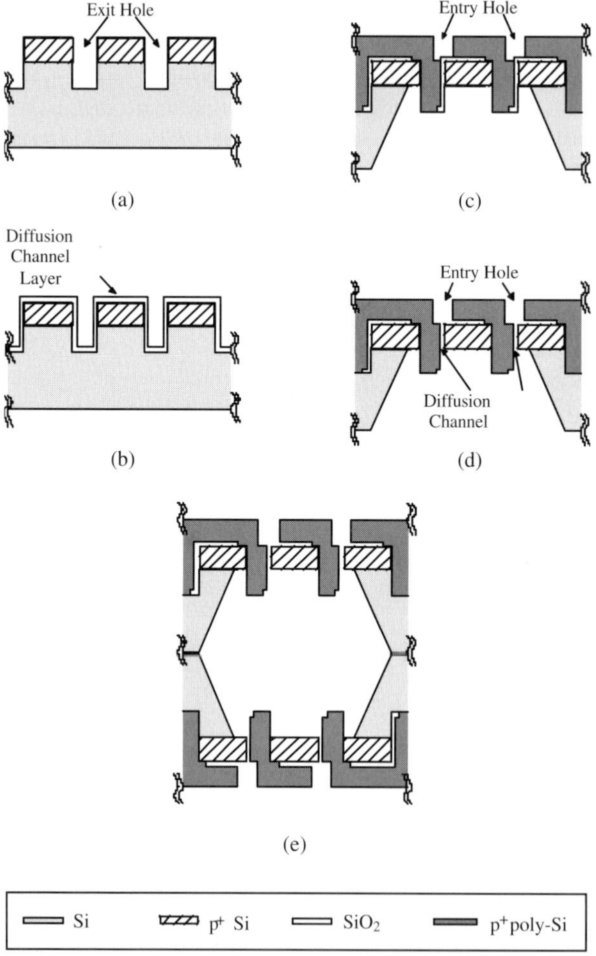

Fig. 4. Essential fabrication steps for the microfabricated biocapsule (not to scale).

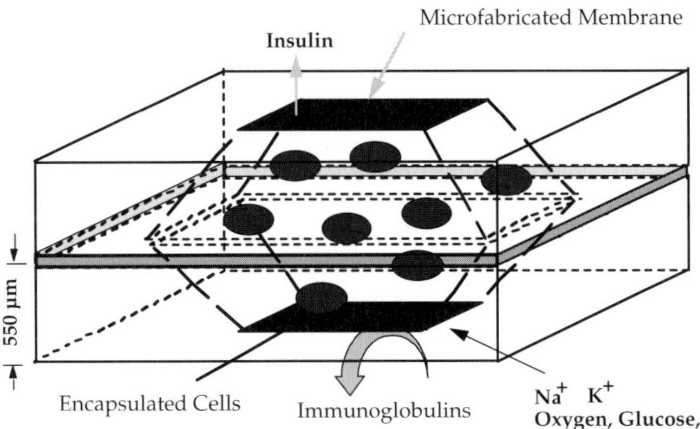

Fig. 5. Schematic of assembled biocapsule consisting of two micromachined membranes bonded together to form a cell-containing cavity bounded by membranes

4. An Opportunity: Intravascular, Systemic Treatment of Metastatic Disease

4.1 Introduction

The stage is set for nanotechnology and microfabrication to be applied to create a new generation of "intelligent" injectable Drug Delivery Systems (DDS). Indeed, the flexibility and precision provided by microfabrication makes it likely that such methods will be applied to produce drug-containing microparticles which are engineered to deliver encapsulated drugs to specific sites of pathology such as tumors. Such particles might be sent on a "seek and destroy" mission: to find and kill cancer cells, for example. In this section we present a hypothetical microfabricated system for treating solid tumors that combines an understanding of tumor biology, current knowledge of the distribution and kinetics of injected particulate systems and the potential of microfabrication to create ligand-targeted, drug-containing microparticles. The biological paradigm is the Natural Killer (NK) cell. This type of white blood cell is an important component of the immune surveillance system that protects all of us from uninvited cells. Indeed, NK cells move through the body looking for abnormal cells. NK cells are able to bind specifically to cancer and other abnormal cells and, after doing so, to eject packets of highly toxic molecules. These molecular torpedoes insert into the surface membrane of target cells and kill them. We believe that the emerging area of nanotechnology provides the potential to create a microfabricated artificial NK cell.

4.2 The Target: Tumor Blood Vessels not Tumor Cells

Primary tumors are generally not the principal cause of morbidity and mortality among cancer victims. Indeed, highly effective treatments exist for treating the majority of primary lesions, including surgery, radiation and focused chemotherapy. Metastatic disease presents a far greater treatment challenge as metastatic lesions typically form in multiple sites and deep in vital organs (Liotta 1993). Surgery is only partially effective and radiation exposure to multiple sites in the body can lead to unacceptable systemic toxicity well before the disease process is under control. Systemic chemotherapy using cytotoxic or biological agents is the only treatment option for many patients with advanced metastatic cancer. Many tumors respond to initial courses of chemotherapy. Unfortunately, after exposure to multiple courses and drugs, cancer cells become resistant and fail to respond to further therapy. Multiple drug resistance (or MDR) is one of the most vexing problems faced by medical oncologists, leading to treatment failures in the vast majority of patients with metastatic disease (Morrow 1993).

Metastatic tumors develop when a small number of cells (or clumps of cells) detach from primary tumors, enter and move through blood vessels or lymphatics, invade tissues at distant anatomical sites and form metastatic foci. The *in situ* proliferation of such cells, and formation of secondary micrometastatic lesions, must rely on nutrients to be provided by normal blood vessels supplying the area. Once the tumors have grown to a few mm^3 in volume, normal vessels are insufficient to support further tumor growth. Further growth of metastatic tumors is supported by factors secreted by the tumor cells themselves, which, in a coordinated fashion, cause new blood vessels to sprout from existing ones. This process, known as angiogenesis, produces a network of blood vessels that supply nutrients to the growing tumor mass and provide yet another avenue for spread of the disease (Holash et al. 1999).

During angiogenesis tumor cells secrete growth factors such as vascular endothelial growth factor (VEGF) and basic fibroblast growth factor (FGFb), both of which stimulate the endothelial cells of normal blood vessels to proliferate and mobilize. The tumor cells also elaborate enzymes, metalloprotinases, which act by carving out tiny channels in the tissue matrix into which the endothelial cells migrate, eventually forming closed tubes (capillaries). Migration of endothelial cells is guided by the interaction of integrins expressed on such cells during proliferating with RGD-containing matrix proteins such as fibronectin.

Angiogenic blood vessels grow in a haphazard pattern and can be quite leaky. Indeed such blood vessels contain many defects and gaps in the endothelium. Intravenously injected particles up to 600 nm in diameter have been shown to extravasate (physically pass from the lumen of the vessel into the interstitium) in tumors implanted in rodents (Yuan et al. 1994). Regrettably, this size window excludes reservoir-containing microfabricated particles. Although such particles can be made in sizes that are safe for IV injection (i.e., in the 5-10 micron range (Conhaim and Rodenkirch 1998; Latres et al. 1992), these are not small enough to permit penetration into the tumor interstitium (i.e., < 600 nm) as can be seen from Fig. 6.

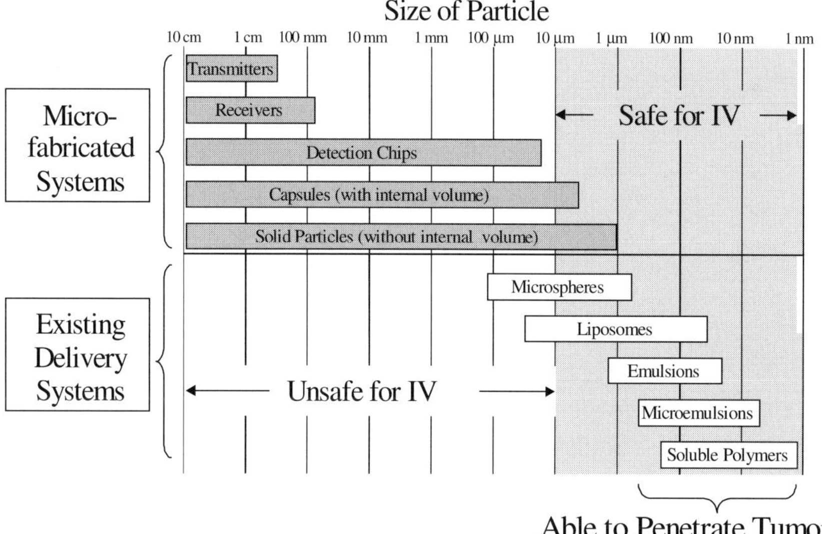

Fig. 6. Relationship between particle size and safety for intravenous administration and tumor penetrability

One strategy for preventing or treating metastatic disease, which dose not require that the particle extravasate into the tumor interstitium (an approach which is gaining favor among oncologists), is to intervene in the process of angiogenesis by destroying the blood vessels which supply tumor cells rather than the tumor cells themselves (Giavazzi and Giulia 1999). By preventing new blood vessels from sprouting, or killing the endothelial cells that form existing tumor capillaries, metastatic tumors may be prevented from growing and spreading. Moreover, no resistance would be expected to develop to such therapy. A blood supply is essential for tumor growth and without it tumors would regress.

4.3 The Carrier: Microfabricated "Nanoparticles"

"Top-down" fabrication of micro-devices, using techniques perfected by the electronics industry, provides the means to create microscopic particles with a unique combination of structural features useful for delivery of cytotoxic drugs to the microvasculature of tumors (Davis 1997). Such particles can be made with extremely precise sizes and shapes and can contain pores, which can act as reservoirs enabling the particle to transport chemotherapeutic drugs. Moreover, the particles may be asymmetrical. For example, the pores or reservoirs can be made to open only to the top face of the particle. The top face (containing the pore openings) can also be chemically modified to contain reactive chemical groups such as primary amino or thiol groups, which can be used to chemically graft protein or other types of ligands to this face only. As will become evident in the discussions below, the unique geometry provided by such microfabrication methods is useful to create the particles proposed here.

Figure 7 exemplifies the type of nanostructure envisioned here. In its longest dimension the particle is about 10 μm. The cytolytic drug is contained within a pore fabricated into the structural material of the particle. After filling, the opening of the pore is plugged with a layer of erodible material such as gelatin. A layer of ligand molecules is chemically grafted to the face of the particle via reactive chemical groups.

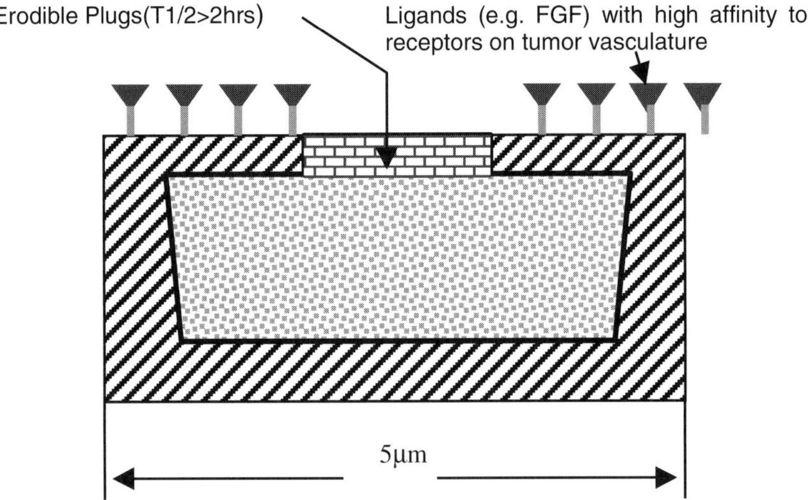

Fig. 7. Microfabricated artificial "killer" cell

4.4 Microfabrication Schemes

The so-called "top-down" approach employs a combination of thin film deposition methods plus photolithography, photoablation and etching techniques to deposit and mold sequential layers of materials on a substrate (Fujimasa 1993). Using this approach materials such as SiO_2 and polymers are applied as a thin film to a sacrificial layer by standard techniques including chemical vapor deposition, sputtering or the like. Subsequent layers of materials may be added, including photoresist materials which, when exposed to the proper wavelength of UV light or other source of irradiation, either undergo a chemical change which renders the resist layer susceptible to the action of etchants (positive resist) or resistant to the action of etchants (negative resist). Photomasks are used to expose only selected areas of such resists to the source of irradiation and etchants are used to dissolve susceptible areas.

4.5 The Targeting Mechanism: Ligands which Recognize the Vascular Endothelium of Tumors

The face of the microfabricated particles to which the pores or reservoirs open may be chemically modified using techniques such as plasma discharge in selected gases to introduce reactive chemical groups (such as primary amines or thiols) or by treatment with appropriate sialylation reagents. For example, the surface of the microparticle may be treated with a silanization reagent to introduce a high density of reactive amino groups onto the silicon surface. Next the surface is treated with an excess of a heterobifunctional reagent (such as SMCC) to render the surface thiol-reactive (Nashat et al. 1998; Zhang et al. 1998). Using this approach these ligands have been chemically coupled to liposomes and viruses and shown to redirect the particle tropism toward cells that express specific receptors (Gu et al. 1999; Rogers et al. 1997). The coupling method insures proper orientation of the ligand on the surface of the microparticles and high ligand density. Retention of ligand specificity and full binding activity is expected. These approaches can be used to graft a layer of ligands to the face of the particles. Suitable ligands are listed in Table 1 and the schematic process of ligand grafting is shown in Fig. 8. Preferred ligands include those that bind to receptors expressed by proliferating vascular endothelial cells and/or tumor cells. One preferred class of ligands suitable for the present invention is growth factors such as FGF, EGF, VEGF and her2 that bind with high affinity to receptors overexpressed on proliferating endothelial or tumor cells. Other suitable ligands include folate and tumor-targeting RGD cyclic peptides that bind $\alpha_v\beta_3$ integrins.

4.6 The Killing Mechanism: Cytolytic Agents (Lysins) Which Act at Short Distances

Natural killer (NK) cells are able to recognize, bind and kill cells infected by certain viruses and cancer cells. Binding is mediated by specific ligand-receptor interactions. For example, NK cells are able to bind to immune complexes which form on the surface of tumor or virally-infected cells after an antibody (elicited by the body's immune system in response to such abnormal cells) binds to specific cell surface determinants which are expressed as a consequence of the disease process. After NK cell binding, intracellular vesicles or granules packed with high concentrations of specific cytolytic molecules move to the cell surface membrane and fuse with it (a process referred to as degranulation), releasing potent packets of these cytolytic agents (much like little torpedoes) (Andersson et al. 1996). If the NK cell is in sufficiently close juxtaposition to a target cell (i.e., within a few nanometers), these lytic agents are able to move through the thin aqueous space separating the two cells and enter the target cell membrane. Entry of as few as 10^6 of such molecules will cause destabilization of the target cell membrane and colloid osmotic lysis and cell death rapidly ensues. Those lysin molecules that do not find and enter a membrane structure within a few nanometers from the site of their release on the surface of the NK cell bind to plasma components such as albumin and their lytic activity is neutralized. In this fashion collateral damage to nearly normal cells is minimized (Fig. 9).

Table 1. Partial list of ligands

Ligand	Receptor	Cell Type
Folate	Folate receptor	Epithelial carcinomas, bone marrow cells
Apolipoproteins	LDL	Vascular endothelial cells
Transferrin	Transferrin receptor	Endothelial cells
Sialyl-Lewis^x	E, P selectin (ELAM-1)	Vascular endothelial cells
VEGF	Flk-1,2	Vascular endothelial cells, tumor cells
Basic FGF	FGF receptor	Vascular endothelial cells, tumor cells
EGF	EGF receptor	Vascular endothelial cells, tumor cells
VCAM-1	$\alpha_4\beta_1$ integrin	Vascular endothelial cells
ICAM-1	$\alpha_1\beta_1$ integrin	Vascular endothelial cells
PECAM-1	$\alpha_v\beta_1$ integrin	Vascular endothelial cells
Osteopontin	$\alpha_v\beta_1$ integrin $\alpha_v\beta_5$ integrin	Vascular endothelial cells
RGD sequences of matrix proteins	$\alpha_v\beta_5$ integrin	Vascular endothelial cells, tumor cells
Anti-c-erbB2 antibody, Fab Fragments of single chain antibody (Fv)	c-erb**B**-2 receptor	Tumor cells

Cytolytic molecules that act only at short distances (such as those found in NK cells) are a preferred class of cytotoxic agent for encapsulation in the proposed particles (Table 2). Indeed, in the context of this approach the microparticles are designed to mimic the behavior of NK cells. A potent cytolytic agent is encapsulated in the internal reservoir of the particles to mimic the natural cytolysins of NK cells. A preferred agent is melittin extracted from bee venom.

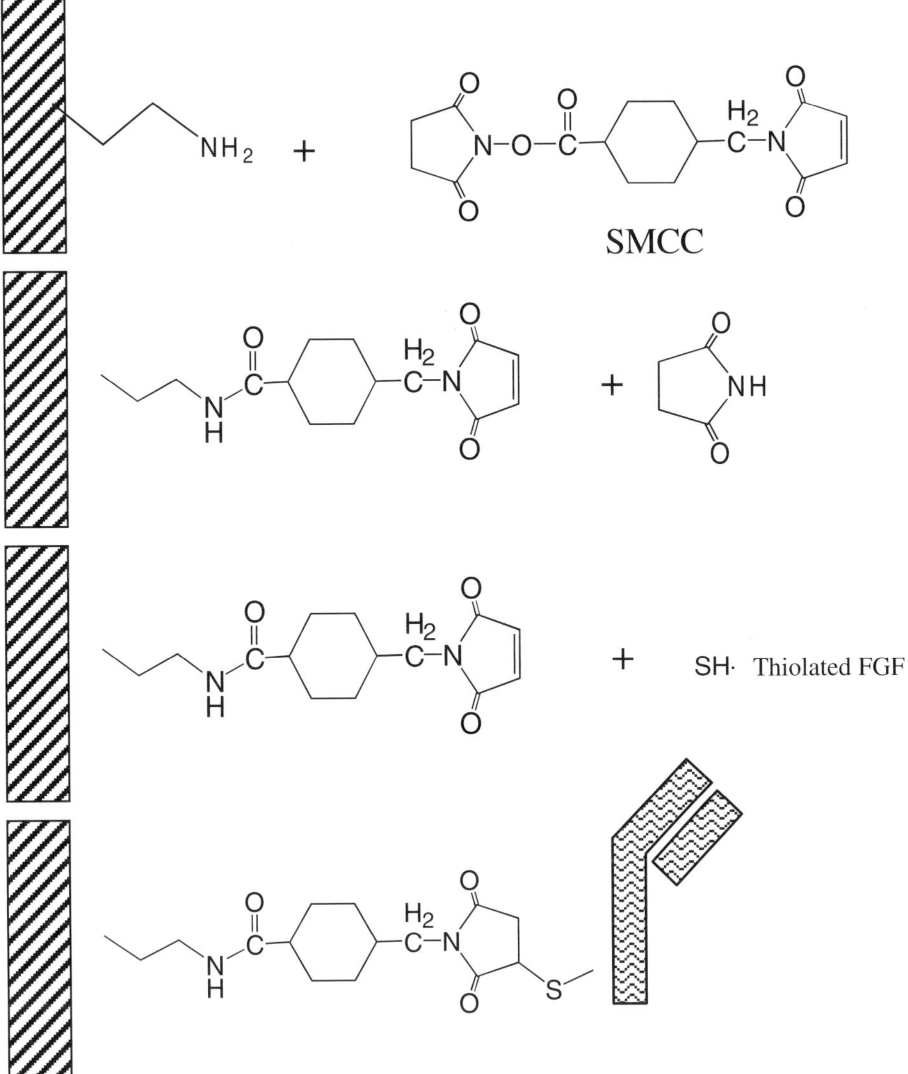

Fig. 8. Grafting of ligands to microparticles

Table 2. Lytic peptides for encapsulation in artificial NK cells

Name of Lysin	Source	Comments
Melittin	*Apis mellifera* (honey bee)	26 amino acid peptide: GIGAVLKVLTTGLPALISWIK RKRQQ-NH$_2$
Paradaxin	*Pardachirus marmoratus* (Red Sea Moses sole)	Short peptide
Hemolysin	*Staphylococcus aureus*	Short peptide
Amoebapore	*Entamoeba histolytica*	77 amino acid residues arranged in 4 alpha-helical domains
Pilosulin	*Myrmecia pilosula* (jumper ant)	56- amino acid residue polypeptide
Magainin	*Xenopus laevis* (skin)	Short peptide
Lentivirus lytic peptide	HIV-1	Carboxy-terminal 29 amino acid residues of transmembrane glycoprotein
NK-lysin	Cytotoxic T lymphocytes	
Perforin	Natural Killer lymphocytes (NK)	N-terminal 22-residue domain

The pores (which connect the face of the particle with the melittin reservoir) are plugged with a material designed to erode in 1-48 hours after IV injection of the particle (such as starches or gelatin). After injection the particles circulate within the bloodstream for several minutes to several hours after which time they are removed by the body's immune system (RES). This time in the circulation is sufficient to bring some of the circulating particles in close contact with the surfaces of proliferating vascular endothelial cells in systemic tumors (including micrometastatic lesions). The face of the particle containing the pore openings has been grafted with a layer of FGF molecules which, after injection, act as high affinity ligands to bind the pore-face of the particle to the surface membrane of proliferating vascular endothelial cells (i.e., those forming new vessels in tumors) which express a high density of FGF receptors. Once bound to the endothelial cell, over the ensuing several hours to several days the plug material erodes and the melittin is released. Importantly, as in the case of tumor cell killing by NK cells, the lysin is released in sufficiently close proximity to the target cell surface membrane to which the particle is bound to allow it to enter the target cell's membrane. Entry of as few as 10^6-10^7 molecules of melittin causes lysis and death of the target endothelial cell. Melittin molecules that do not enter the juxtaposed target cell membrane, and melittin molecules released from particle elsewhere in the body (i.e. those that have not bound to endothelial target cells), are inactivated by binding to albumin and thus do not cause toxicity to normal cells.

IV-MEDDS

1. Binding (a few hours after injection)

2. Plug rupture, drug release (several hours)

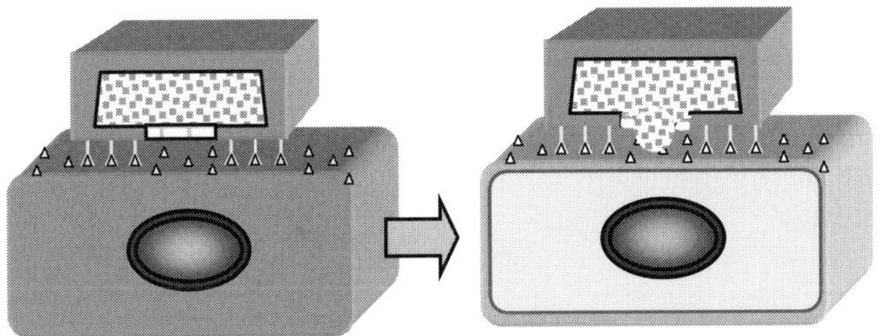

3. Pore formation-cell lysis and death (12-48 hours)

Fig. 9. Mechanism of artificial NK cell action

5. Prospects

Although the microfabricated particles proposed here represent only a theoretical possibility at this point, the prospects for creating such a system are not far fetched provided that certain technical obstacles can be overcome. The most serious challenges include demonstrating the safety of such particles injected intravenously, maintaining sufficiently long circulation time to reach and penetrate disease-site targets and engineering techniques to load drugs and release them at the proper time and place. Nanotechnologists are now turning their attention to the problems of drug delivery and the pace of technological innovation in the area is accelerating. Moreover, the instincts of nanoengineers and pharmaceutical scientists are now converging on the notion that it is only a matter time before nanotechnology finds a significant role to play in drug delivery.

References

Aebischer P, Goddard M, Signore AP and Timpson RL (1994) Exp Neurol 126(2): 151-158

Akin T and Najafi K (1994) IEEE Transactions on Biomedical Engineering 4(4): 305-313

Anderson DJ, Najafi K, Tanghe SJ, Evans DA, Levy KL, Hetre JF, Xue X, Zappia JJ and Wise KD (1989) IEEE Transactions on Biomedical Engineering 36(7): 693-704

Andersson M, Gunne H, Agerberth B, Boman A, Bergman T, Olsson B, Dagerlind A, Wigzell H, Oman HGB and Gudmundsson GH (1996) Vet Immunol Immunopathol 123-126

Baxter GT, Bousse LJ, Dawes TD, Libby JM, Modlin DN, Owick JC and Parce JW (1994) Clinical Chemistry 40(9): 1800-1804

Benjamin JD, Mears AL and White JC (1988) Active silicon implant devices. United States Patent No. 4,793,825

Bergveld P (2000) Biomed Microdevices 185-195

Brazzle JD, Papautsky I and Frazier AB (1998) Proc. SPIE Micro Fluidic Devices and Systems 116-124

Campbell PK, Jones KE, HR J, Horch KW and Nomann RA (1991) IEEE Transactions on Biomedical Engineering 38(8): 758-768

Chen J and Wise K (1994) IEEE Solid-State Sensor and Actuator Workshop, Hilton Head, SC

Chu W-H and Ferrari M (1998) Micromachined Capsules Having Porous Membranes and Bulk Supports, United States Patent No. 5,798,042

Colton CK (1996) Trends Biotechnol 158-162

Conhaim RL and Rodenkirch LA (1998) J Appl Physiol 85(1): 47-52

Davis SS (1997) Trends Biotechnol 217-224

Deamer DW and Akeson M (2000) Trends Biotechnol 147-151

Desai TA, Chu WH, Tu JK, Beattie GM, Hayek A and Ferrari M (1998) Biotechnol Bioeng 118-120

Desai TA, Hansford D and Ferrari M (1999a) J Membr Sci 221-231

Desai TA, Hansford DJ, Kulinsky L, Nashat AH, Rasi G, Tu J, Wang Y, Zhang M and Ferrari M (1999b) Biomed Microdevices 11-40

Edell DJ (1986) IEEE Transactions on Biomedical Engineering 33(2): 203-214

Edell DJ, Churchill JN and Gourley IM (1982) Biomat Med Dev Artif Org 10(2): 103-122

Evans J, Liepmann D and Pisano AP (1997) IEEE Proc Micro Electro Mechanical Systems Workshop, Nagoya, Japan

Ferrari M (2000) Therapeutic Microdevices and Methods of Making and Using Same, United States Patent No 6,107,102, (issued).

Ferrari M, Chu W-H, Desai T, Hansford D, Mazzoni G, Huen T and Zhang M, Silicon Nanotechnology for Biofiltration and Immunoisolated Cell Xenografts, *Thin Films and Surfaces for Bioactivity and Biomedical Applications*, MRS, Vol. 414, Eds. Catherine Cottell, et al, pp. 101-106, Materials Research Society 1996.

Fodor SPA (Washington DC) (1997) Science 393, 395

Fujimasa I Micromachining technology and biomedical engineering (1993) Appl Bio-chem Biotechnol 38(3):233-42

Giavazzi R and Giulia T (1999) Forum (Genova) 9: 261-272

Gourley PL (1996) Nature Medicine 2(8): 942-944

Gu D-L, Gonzalez AM, Printz MA, Doukas J, Ying W, D' Andrea M, Hoganson DK, Curiel DT, Douglas JT, Sosnowski BA, Baird A, Aukerman SL and Pierce GF (1999) Cancer Res 2608-2614

Hansford D, Desai T and Ferrari M, "Nano-Scale Size-Based Biomolecular Separation Technology", Biochip Technologies, Eds. Kricka and Cheng, Harwood Academic Publishers, pp. 341-362, 2001.

Henry S, McAllister DV, Allen MG and Prausnitz MR (1998) J Pharm Sci 922-925

Henry S (1997) Master's Thesis: Department of Chemical Engineering. Georgia Insti-tute of Technology: Atlanta, GA

Holash J, Wiegand SJ and Yancopoulos GD (1999) Oncogene 5356-5362

Kittilsland G, Stemme G and Norden B (1990) Sensors and Actuators A21-A23: 904-907

Kovacs GTA, Storment CW, Halks-Miller M, Belczynski CR, Della Santina CC, Lewis ER and MNI (1994) IEEE Transactions on Biomedical Engineering 41: 567-577

Latres E, Closa D, Gomez-Sierra JM, Alemany M and Remesae X (1992) Arch Int'l Physiol Biochim Biophys 100(3): 263-265

Lim F and Sun AM (Washington DC) (1980) Science 90:8-10

Lin L and Pisano AP (1999) IEEE J Microelectromechanical Systems 8(1): 78-84

Li RH (1998) Adv Drug Delivery Rev 87-109

Liotta LaS-S (1993) Principles of Molecular Cell Biology of Cancer: Cancer Metas-tasis in Cancer Principles and Practice In: DeVita V, Hellman S and Rosenberg S (eds) Lippincott: Philadelphia, pp134-139

Lonsdale HK and Membr J (1982) Sci 81-181

McAllister DV, Henry S, Allen MG and Prausnitz MR (1998) Proc Int Symp Con-trolled Release Bioact Mater 30-31

McConnell HN, Owicki JC, Parce JW, Miller DW, Baxter GT and Wada HG (1992) Science 257(5078): 1906-1912

McGall GH, Barone AD, Beecher JE, Diggelman M, Fodor SPA, Goldberg MJ, Ngo N and Rava RP (1999) Light-directed synthesis of high-density oligonucleotide probe arrays for nucleic acid sequence analysis In: R. Epton (Ed), Innovation and Perspectives in Solid Phase Synthesis & Combinatorial Libraries 97-100 Mayflower Scientific Limited, Birmingham

Morrow CaC (1993) Mechanisms of Antineoplastic Drug Resistance in Cancer Prin-ciples and Practice of Oncology In: DeVita V, Hellman S, and Rosenberg S (eds) Lippincott: Philadelphia, pp340-348

Nashat AH, Moronne M and Ferrari M (1998) Biotechnol Bioeng 137-146

Reed ML, Wu C, Kneller J, Watkins S, Vorp DA, Nadeem A, Weiss LE, Rebello K, Mescher M, Smith AJC, Rosenblum W and Feldman MD (1998) J Pharm Sci 1387-1394

Roberts T, De Boni U and Sefton MV (1996) Biomaterials 267-75

Rogers BE, Douglas JT, Ahlem C, Buchsbaum DJ, Frincke J and Curiel DT (1997) Gene Ther 1387-1392

Sagen J, Wang H, Tresco PA and Aebischer (1993) J Neurosci 2415-23

Santini JT, Jr, Cima MJ and Langer R (London) (1999) Nature 335-338.

Talbot NH and Pisano AP (1998) Tech Dig, Solid-State Sensor and Actuator Work-shop Hilton Head, SC

Tu JK, Huen T, Szema R and Ferrari M (1999) Biomed Microdevices 113-119

Van Rijn CJM and Elwenspoek MC (1995) IEEE MEMS 83-87

Van Rijn CJM, Veldhuis GJ and Kuiper S (1998) Nanotechnology 9(4): 343-345

Volkmuth WD, Duke T, Austin RH and Cox EC (1995) Trapping of branched DNA in microfabricated structures (1992) Proc Natl Acad Sci (USA) 92(15): 6887-6891

Winn SR, Hammang JP, Emerich DF, Lee A, Palmiter RD and Baetge EE (1994) Proc. Natl Acad Sci (USA) 2324-8

Wise KD, Najafi K, Ji J, Hetke JF, Hoogerwerf A, Anderson DJ, BeMent SL, Ghazzi M, Baer W, Hull T, and Yang Y (1990) Annual Conference of the IEEE Engineering in Medicine and Biology Society, (Philadelphia) 12(5): 2334-2335

Yang X, Yang JM, Wang XQ, Meng E, Tai YC and Ho CM (1998) Proceedings of IEEE The Eleventh Annual International Workshop on Micro Electro Mechanical Systems (MEMS '98), Heidelberg, Germany, January 25-29, 1998

Yuan F, Leunig M, Huang SK, Berk DA, Papahadjopoulos D and Jain RK (1994) Cancer Res 335:2-6

Zhang M, Desai T and Ferrari M (1998) Biomaterials 953-960

Zhang M and Ferrari M (1997) Biotechnol Bioeng 618-625

Photobonding of Biomolecules for Biochip Manufacturing

Gao H., Juvet M., Karlen Y., Menotti E. and Sigrist H.

Abstract

Immobilization of biomolecules on material surfaces while retaining their biological functions is a major challenge in biochip manufacturing. Photobonding is a unique procedure for bioprobe immobilization on "inert" surfaces by a light-induced process. We have designed and synthesised photolinker polymers carrying several photoactivatable chemical functional groups (e.g. aryldiazirines). When irradiated with light, the photoactivatable chemical groups turn into reactive intermediates. Generated intermediates bond to all neighboring materials that are either substrate or biomolecule. In biochip technology, hydrophilic polysaccharides (or bovine serum albumin) are appreciated as molecular barriers due to their biological inertness and biocompatibility.

In this study dextran was modified with photoactivatable groups. The product, OptoDex, served as the basis for the syntheseis of several photoactivatable Opto-Dex analogues. Photolinker polymer mediated covalent immobilization of biomolecules was attained under mild and non-invasive conditions. Neither reactive functional groups on the biomolecule nor on the material surface were required to achieve single step probe molecule immobilization. The described photobonding process combines several unique properties: local addressability, light-dependent covalent bonding, single-step multicompent immobilization and chemical multifunctionality through modification of the parent molecules. The construction of biochip microarrays was achieved by combining ink-jet printing and photobonding technologies.

1. Introduction

Biochips are micro-scale devices preferably made by methods derived from semiconductor technologies. They are highly miniaturized analytical devices exhibiting massive parallelism of function. Fodor and collaborators (Fodor et al. 1990; Jacobs and Fodor 1994) have pioneered the use of photolithography for on-chip synthesis of highly dense arrays of oligonucleotides and short peptides. To date, a feature size of 10 µm enables the formation of high density arrays. Nanogen specialises in considerably less dense arrays that are made by processes used in microelectronics and incorporate pre-synthesised oligonucleotides (Swanson et al. 2000). Each position of a Nanogen array is addressed electronically in order to facilitate both the hybridisation of targets and the removal of mismatched nucleic acids.

Increasingly unsupportable cost and time constraints gave rise to the science and business of functional genomics: the identification of gene expression patterns and single nucleotide polymorphisms that occur frequently in diseased populations. Whereas major efforts are currently placed on biochips designed for the detection of target nucleic acid binding to DNA or RNA arrays, there is increasing demand for multiplex detection of protein-protein interactions in array systems. In contrast to the engineering of biochip arrays with probing nucleic acids, the manufacturing of protein arrays requires novel technologies and immobilization approaches. Individual proteins differ significantly in molecular stability, surface charge, local surface polarity and many more physical and kinetic parameters. The expression of the individual biological function of proteins on material surfaces is often difficult to achieve, incidentially even for a single type of protein.

In order to circumvent obstacles set by the molecular idiosyncrasies of proteins and biomolecules in general, tools and processes were designed for addressable immobilization of biomolecules including proteins on material surfaces. Light-sensitive photolinker polymers were used to immobilize probing molecules on surfaces and at the same time passivate the surface by suppressing the bonding of non-target molecules. In analogy to biological systems, the polysaccharide dextran and analogues thereof were used for biomolecule (protein, polysaccharide, oligonucleotide) immobilization. The dextran-based photolinker polymer com- bined the advantageous properties of surface active oligosaccharides [2, 7] with local addressability, light-dependent covalent bonding and single-step multicomponent immobilization (Barié et al. 1998; Gao et al. 1994; Sigrist et al. 1995).

2. Experimental Procedures

2.1   Materials and Methods

Dextrans were purchased from Molecular Probes (Oregon USA): amino dextran (40 kD); rhodamine labeled-dextran. The immunochemicals mouse IgG, goat anti-mouse IgG and fluorescence-labeled antibodies were from Pierce, synthetic oligonucleotides were from Microsynth, Balgach, Switzerland and 3'-terminal desoxynucleotidyl transferase (3'TDT) was purchased from Amersham- Pharmacia. Chemicals: 1-ethyl-3-(3-dimethylaminopropyl) carbo-diimide (EDC), ethanolamine, N-hydroxysuccinimide (NHS) and glutaric anhydride were purchased from Fluka, Pierce or Sigma. 3-(trifluoromethyl)-3-(m-isothiocyano- phenyl) diazirine (TRIMID) was synthesized according to Dolder et al. [3]. TRIMID-derivatized dextran (Optodex) and its analogues (OptoDex B, OptoDex C) were synthesized according to Gao et al. (manuscript in preparation). Phosphate buffered saline (PBS) buffer contained 50 mM sodium phosphate and 150 mM NaCl. The pH was adjusted to 7.4. PBS containing 0.05% (v/v) Tween 20 was used as washing buffer. OPA buffer (One-Phor-All) contains tris-acetate, 10 mM, magnesium acetate, 10 mM and potassium acetate, 50 mM, pH 6.9. A sodium acetate buffer (10 mM, pH 5.0) was used for the amidation reaction of proteins with NHS/EDC activated OptoDex C. Ethanolamine, 1.0 M, was adjusted pH to 8.5 and used for capping of NHS/EDC activated OptoDex C surfaces.

Antibody (protein) concentrations were quantitated by applying the extinction coefficient E 280 (1% solution) = 14.0.

Optical grade silicon nitride (Si_3N_4) on silicon (manufactured by CSEM) and alkali free glass AF 45 (Schott & Schleifer) were used as substrates.

An Oriel Lamp (350 nm, 11 mW/cm^2) and a Stratalinker 2400 UV source (365 nm; 0.9 mW/cm^2) were used for photoactivation.

Surface fluorescence was detected with the Axiovert 100 microscope (Zeiss, Switzerland). A fluorescence waveguide sensor (Voirin et al. 1999) was used to monitor binding of oligonucleotides after 3' end labeling with the enzyme 3' termial transferase and a fluorescent desoxynucleotide triphospate (Cy5-desoxy-cytosin triphosphate). The instrument allows the detection of fluorescent probes in the evanescent field. For excitation of the fluorescent labels laser light is coupled into the waveguide via a grating coupler. The evanescent wave of the guided mode excites the fluorescent labels bound to or associated with the waveguide surface. For detection emitted fluorescent light is also guided along the waveguide and coupled out by the same grating coupler. The instrument has been used in a dual sample reading mode.

2.2 Single Step Co-immobilization of Biomolecules

2.2.1 Immobilization of Antibodies

Glass and silicon nitride chips (12 x 12 mm) were pre-treated over night with concentrated HCl, extensively washed with bidistilled H_2O and dried. OptoDex-A was mixed with IgG (or fluorescence-labeled IgG) in 1:100 diluted PBS buffer. Nanoliter volumes were arrayed on glass or Si_3N_4 chips using a piezoelectric micropipette (Nanoplotter, GeSim). After drying under vacuum (5.0 x 10^{-2} mbar) for 3 hrs chips were irradiated for 4 min with the Oriel light source, then washed with PBS buffer containing 0.02% Tween 20 (3 x 0.5 ml), PBS buffer (3 x 0.5 ml) and bidistilled water (3 x 0.5 ml).

Immobilized IgG was detected by fluorescence microscopy either directly, for detection of photobonded FITC-IgG, or after immunocomplexation of photobonded IgG with fluorescence-labeled antibodies.

2.2.2 Immobilisation of Fluorescent Dextran on Glass and Si_3N_4 Surfaces

OptoDex-B was mixed with rhodamine-dextran. Procedures described in Sect. 2.2.1 were applied for arraying, photoimmobilization, washing and rhodamine fluorescence detection by microscopy.

2.2.3 Immobilization of Oligonucleotides on Si_3N_4 Waveguide Chips

OptoDex-B was mixed with oligonucleotides at a weight ratio 1:1 and applied to Si_3N_4 waveguide chips at the final OptoDex-B density of 240 ng/mm^2. Following drying and photoimmobilization as described in 2.2.1, the chips were washed with KSCN, 3.0 M (3 x 0.5 ml), PBS/Tween (3 x 0.5 ml) and PBS buffer (3 x 0.5 ml). Each washing step included shaking for 5 min. Tailing of the immobilized oligonucleotide was as recommended by the supplier of the 3' terminal deoxynucleotidy transferase (3'TdT): 3'TdT (1.0 µL, 24 IU/µl) was mixed with dATP (1.0 µL, 10 mM) and Cy5-dCTP (0.5 µL, 1.0 mM) in a final volume of 10 µl OPA buffer. The mixture was applied to each chip and the chips were incubated for 30 min at 33°C. The enzymatic tailing reaction was stopped by washing the surface once with KSCN, 3.0 M, and three times with PBS/Tween. Cy5 fluorescence retained on the chips was determined with the fluorescence waveguide sensor system.

2.3 NHS/EDC Mediated Immobilization of Human IgG to Photobonded OptoDex C on Glass or Si_3N_4

OptoDex C (prepared according to Gao et al. 2000, manuscript in preparation) was dissolved in 1:100 diluted PBS buffer (2.0 mg/ml) and spotted with a pipette onto glass or *Si_3N_4* chips (2.0 µl/spot). After drying for 3 h at ambient temperature under vacuum the chips were irradiated for 4 min with the Oriel lamp. Regular patterns of differing dimensions were obtained by irradiation through a microstructured mask (chromium on glass) in contact mode. Chips were washed with KSCN, 3.0 M (3 times), PBS containing 0.02% Tween 20 (3 times) and bidistilled water (3 times). If required, washed chips were stored at 4°C till use. Washed chips were placed in Falcon plate wells. EDC (400 mM) and NHS (100 mM) were prepared freshly with bidistilled water. Equal volumes of two solutions were mixed thoroughly and the solvent mixture was applied (10 µl/chip). Following incubation for 1 h at ambient temperature, the reaction medium was removed and the chips were rinsed with 10 mM sodium acetate buffer, pH 5.0. Human IgG was dissolved in acetate buffer at a concentration of 100 µg/ml and 10 µl of the solution was applied to each chip. The amidation reaction was carried out for 1 h at ambient temperature. Modified chips were then washed with PBS/Tween 0.02% (3 times) and PBS buffer (3 times). Immobilized IgG was detected by fluorescence microscopy after immunocomplexation with fluorescence-labeled anti human IgG.

3. Results and Discussion

3.1 Photobonding

Following basic concepts and protocols of preceding work on light-dependent immobilization of biomolecules to material surfaces (Barié et al. 1998; Gao et al. 1994; Sigrist et al. 1995), the dextran-based photolinker polymer OptoDex was used for covalent biomolecule binding. Essential chemical functions of the polymer are depicted in Fig. 1A. The polysaccharide chain of dextran contains statistically distributed substitutions with the aryldiazirine as photoactivatable unit and statistically occurring hetero- functions X (X= H: Optodex A; X= acetyl: OptoDex B; X= amidoglutarate: OptoDex C). The degree of substitution with either substituent was 5 to 8 mol per mol dextran. In accordance with the physical properties of the parent photoreagent TRIMID (Dolder et al. 1990), the aryldiazirine absorbs 350 nm light (Fig. 1B). The absorption band does not coincide with the absorption of most biomolecules. 350 nm light was therefore used to activate the photoreactive diazirine. Photolysis generates highly reactive carbenes which are known to preferably insert into C-H, C-C, C=C, N-H, O-H or S-H bonds. As a consequence covalent chemical bonds were formed between target substrates, the photolinker polymer and biomolecules. In agreement with our previous investigations using aryldiazirine modified bovine serum albumin as photolinker polymer (Barié et al. 1998; Gao et al. 1994; Sigrist et al. 1995) it was found that photobonding with OptoDex retained biological functions such as ligand binding, enzymatic activities, immunocomplexation and nucleic acid base-pairing.

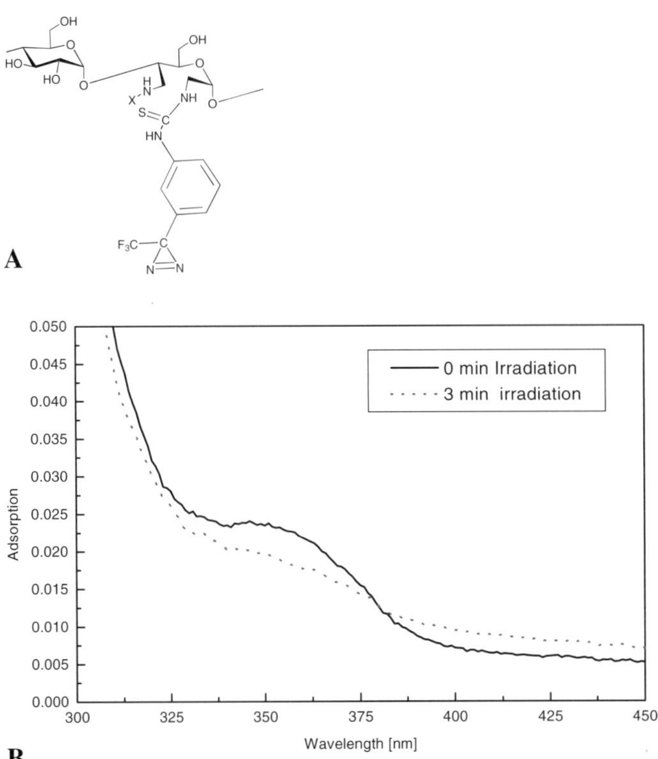

Fig. 1. The chemical structure of Optodex (**A**) and photoactivation of OptoDex (**B**) The absorption spectrum of OptoDex A in aqueous medium is recorded before and after exposure to activating light

3.2 Optodex Mediated Single-step Binding of Biomolecules to Material Surfaces

OptoDex mediated light-dependent immobilization of biomolecules to material surfaces was achieved with antibodies, polysaccharides and oligonucleotides as probe molecules. Fluorescence-labeled antibodies were mixed with OptoDex and spotted in a microarray format on glass surfaces using piezoelectric pipettes for the deposition of nL samples in array formats. After photoimmobilization of fluorescence labeled IgG, arrayed fluorescent spots were observed by microscopy (Fig. 2A). Figure 2B shows the result obtained with photobonded human IgG after incubation with fluorescence-labeled goat anti-human IgG. The result indicates successful formation of immunocomplexes with photobonded IgG and demonstrates the feasibility of protein biochip array manufacturing applying OptoDex mediated photoimmobilization procedures.

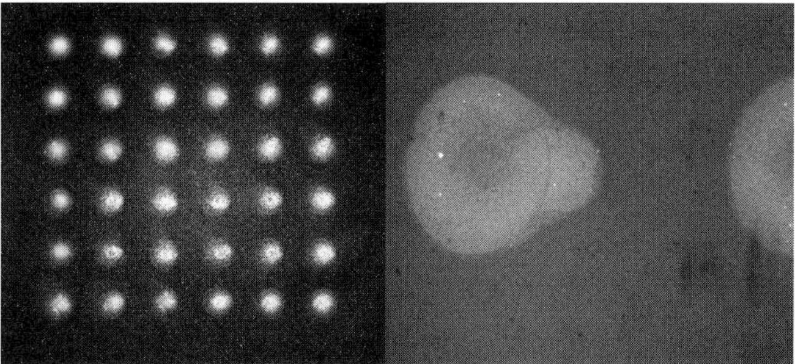

Fig. 2. Microarray of photobonded immunoglobulin. **(A)** FITC-labeled human IgG was mixed with Optodex and printed on glass by ink-jet printing (1 nL per spot, 0.5 mm centre to centre distance, 3 x 3 mm array). Photoimmobilisation and detection procedures are described in the Methods section. **(B)** Following photoimmobilization of human IgG on a glass chip the surface was saturated with 1% BSA and probed with FITC labeled goat anti-human IgG antibodies

Meanwhile pioneering experiments for photobonding and arraying of polysaccharides and oligonucleotides were carried out. Figure 3 shows a microarray of a rhodamine-labeled polysaccharide (dextran) photobonded to a plane Si_3N_4. Similar protocols were applied for single step immobilization of oligonucleotides. Disposable waveguide sensor chips with surface exposed waveguiding Si_3N_4 were used as substrates for photobonding a 40 mer oligonucleotide probe. Binding of the oligonucleotide was mesured with the fluorescence waveguide sensor system (Voirin et al. 1999) after enzymatic oligonucleotide chain extension of the DNA probe with 3' terminal transferase and Cy5 labeled dCTP as enzyme substrate. The experiment substantiates the presence of photobonded oligonucleotides. Successful immobilization depends on light activation. The 3' end of the immobilized probe is accessible to the terminal transferase (Table 1).

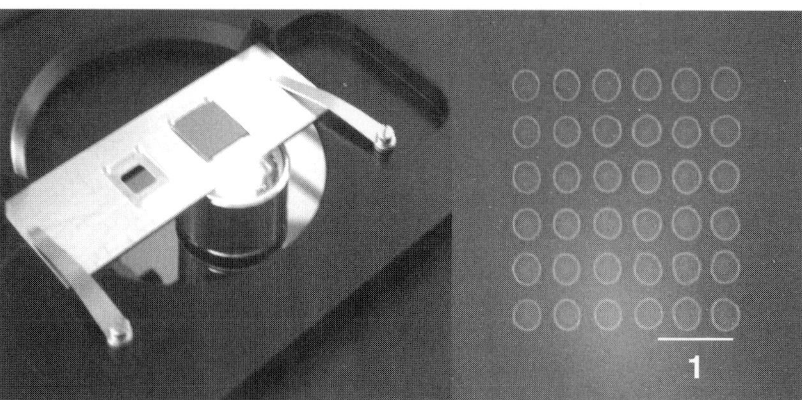

Fig. 3. Microarray of a rhodamine-labeled polysaccharide (dextran) on Si_3N_4 using OptoDex as photolinker polymer

3.3 Covalent Binding of Antibodies to Bonded OptoDex C Surfaces Applying Carbo diimidide Chemistries

Material surfaces modified with carboxyl derivatized OptoDex (OptoDex C) provide spacer extended carboxyl functions for further functionalization. Antibodies as well as other proteins or aminated probe molecules can be covalently linked to the surface by amidation with EDC/NHS activated carboxyl groups of OptoDex C. Following this route of surface functionalization, immobilized IgG was detected after immunocomplexation with fluorescence-labeled anti-IgG antibodies. Figure 4 shows micro-patterns generated by mask-assisted photoprinting of Opto-Dex C. After photoimmobilzation of surface-spread OptoDecx-C and extensive washing, human IgG was covalently linked to surface bonded OptoDex C by amidation. Rhodamine-labeled anti human IgG antibodies were then applied to establish the biopattern shown in Fig. 4 A-C. Target binding intensity, as detected by microscopy, followed the concentration of the antigen applied.

Considering both the biological requirements and device engineering, there are definite advantages in using light for covalent biomolecule immobilization. The successful covalent immobilization of various types of biomolecule including proteins, polysaccharides and oligonucleotides illustrates the versatility and, in conjunction with addressable probe deposition, the potential of photochemical processes for biochip manufacturing.

Table 1. OptoDex mediated photobonding of a 40-mer oligonucleotide

Sample	Light Activation	Oligo- nucleotide	3'TdT extension	Fluorescence (arbitrary units)
Oligonucleotide	+	+	+	39.8 ± 2.6
NO oligonucleotide	+	-	+	4.4 ± 1.3
NO light activation	-	+	+	0.09

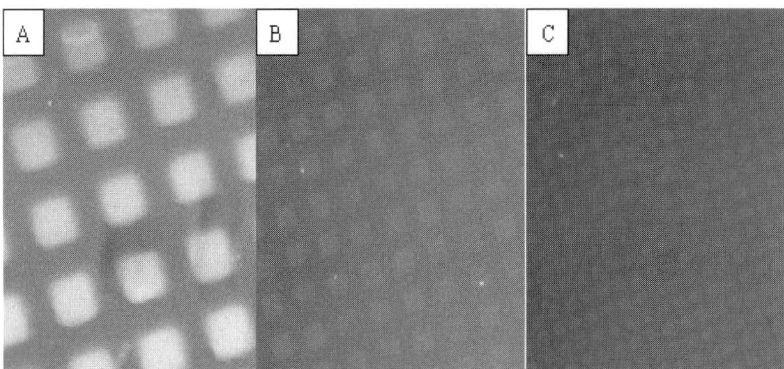

Fig. 4. Mask-assisted photobonding of OptoDex C (carboxylated) onto glass differing in the surface density applied: **A**: 800 ng/mm^2; **B**: 400 ng/mm^2; **C**: 200 ng/mm^2. Human IgG was coupled to photobonded OptoDex C applying carbodiimide chemistries. Surfaces were probed with fluorescence-labeled anti-human IgG. (Pattern sizes: A: 40 x 40μm; B: 20 x 20 μm and C: 10 x 10 μm)

OptoDex was mixed with the oligonucleotide and photoimmobilized on a Si$_3$N$_4$ optical chip. The presence and physical accessibility of the probing oligonucleotide was demonstrated by 3'endlabeling of the photobonded probe with 3' terminal deoxynucleotidyl transferase (3'TDT) and Cy5-dCTP. The extent of oligonucleotide was quantitated with the fluorescence waveguide sensor system

Acknowledgements

The Swiss Priority Programs on Biotechnology, Micro- and Nanosystem Technology and Materials Research are gratefully acknowledged for financial support.

References

Barié N, Rapp M, Sigrist H and Ache HJ (1998) Biosensors and Bioelectronics. 13: 855-860
Bergström J, Löfas S and Johnsson B PCT patent application WO 90/05303
Dolder M, Michel H and Sigrist H (1990) J Prot Chem 9: 407-415
Fodor SPA, Stryer L, Winkler JL, Holmes CW and Solas CW PCT WO 92/10092
Gao H, Kislig E, Oranth N and Sigrist H (1994) Biotech Appl Biochem 20: 251-263
Jacobs JW and Fodor SPA (1994) Trends in Biotechnology 12: 19-26

Petro M, Gemeiner P and Berek D (1994) J Chromatogr A665: 37-45
Sigrist H, Collioud A, Clémence JF, Gao H, Luginbühl R, Sänger M and Sundarababu G (1995) Opt. Eng 34: 2339-2348
Swanson P, Gelbart R, Atlas E, Yang L, Grogan T, Butler WF, Ackley DE and Sheldon E (2000) Sensors and Actuators B64: 22-30
Voirin G, Gehriger D, Parriaux O and Usievich B (1999) Proc. SPIE 3620: 109-116

Opportunities and Obstacles to the Routine Implementation of Microchips

Kricka L. J. and Wilding P.

Abstract

Miniaturization is an important and active area of development in the analytical sciences. Small, portable, inexpensive micro miniature analyzers (microchips, lab-on-a-chip, micro arrays, gene chips, bio electronic chips) are advantageous and have been developed to perform a number of analytical techniques. Despite the opportunities for microchip analyzers in the clinical laboratory, there are still important obstacles associated with their use, such as user-friendly interfaces between a human and a microchip and the level of micro miniaturization that is desirable or practical. This chapter examines the opportunities and obstacles to the routine implementation of microchips, with an emphasis on the clinical and biomedical applications.

1. Introduction

The clinical laboratory has been subject to many forces that have changed its organizational structure, range of diagnostic testing services and the way in which it communicates with the healthcare community. Some of the many factors currently impacting the clinical laboratory include noninvasive testing technologies (e.g., infrared methods), point-of-care testing, information technology (e.g., the world wide web, e-commerce), reimbursement for testing, training and supply of qualified staff, automation and robotics, the human genome project and finally, miniaturization technology. Miniaturization is an important trend in the analytical sciences and is set to have a considerable impact on clinical testing (e.g., increased opportunities for decentralization of testing). This article examines the opportunities and obstacles to the routine implementation of microchips, with an emphasis on the clinical and biomedical applications.

2. Microminiature Analyzers

Microminiaturized analyzers based on an analytical microchip component offer several advantages: they are small and lightweight, and hence portable, and also they have low reagent consumption and require only small volumes of sample (nL-uL) (Cheng and Kricka 20009; Kricka 1998; Wilding P and Kricka 1999). To date many of the conventional analytical techniques: e.g., capillary electrophoresis

(CE)(Colyer et al. 1997; Dolnik et al. 2000; Ueda et al. 2000), mass spectrometry (Wen et al. 2000; Xue et al. 1997), PCR (Belgrader et al. 1998; Cheng et al. 1996; Hofgartner et al. 1999; Kopp et al. 1998; Ross et al. 1998; Waters et al. 1998), immunoassay (Chiem and Harrison 1998)and DNA hybridization (Gerry et al. 2000; Pease et al. 1994) have been adapted to microchip formats (chips constructed from silicon, glass or plastic).

The current range of micro miniature analyzers can be classified into three main types: microchips that provide microfluidic systems (e.g., lab-on-a-chip), micro arrays (e.g., gene chips) and bioelectronic chips (these chips incorporate microelectronic components to enhance the scope of the analytical capabilities of the chip). So-called microchips have also been developed that simply use micromachined silicon chips as substrates on which to mount ion-selective electrodes or devices to measure capacitance. A good example is the cartridge of the i-STAT (www.i-stat.com) device for measurement of electrolytes and blood gases. These chips have been fabricated to support micro electrodes that measure potential differences or capacitance. With the aid of layered reagents over small electrodes, these systems can detect and quantitate a variety of analytes (Wilding and Ciaverelli 2000).

A key advantage of microchip analyzers is integration of all of the steps in an analytical process on the same microchip in a way that is not easily possible with the corresponding macroscale counterparts. The first examples of commercial microchip-based devices combine analytical reactions and detection and separation and detection (e.g., devices for water quality testing, CE analyzers). Prototype microanalyzers with more extensive integration have been exhibited and are expected to be available in the near future. However, despite the considerable opportunities for microchip analyzers, as yet none have been introduced into the clinical laboratory. This is because there are still important obstacles and issues associated with their implementation in a clinical laboratory setting. These are discussed briefly in the following sections.

3. Critique of Microanalyzers for Clinical Testing

A number of issues can be identified that have a direct bearing on the future success of microanalyzers in routine clinical testing.

User-friendly Interface. Microchip devices are small and most are constructed on chips that are approximately 1 cm x 1 cm in area and a few mm thick. At this scale the interface between a human operator and a microchip becomes critical. Consideration must be given to the practicalities of sample and reagent addition via minute entry ports (e.g., 500 um x 500 um) into the microstructures within the microchip and the physical handling of the device. Thus far most manufacturers have opted to incorporate microchips into plastic cassettes or cartridges in order to facilitate handling and to protect the fragile silicon or glass chips.

Level of Micro-Miniaturization. It is clear that there are limits to the level of micro miniaturization that is desirable or practical. There are limits imposed by the need for a human to operate the device and to input sample or reagents. Likewise, there are physical and economic limits determined by the manufacturing process. In addition there are constraints determined by the detection limit of the analytical method in use on the microchip. For example, consider the analysis of a 1 uL, a 1 nL, and a 1 pL sample of a 10 femtomolar solution. At 1 uL there are a total of approximately 6000 molecules in the sample, at 1 nL this falls to 6 molecules, and at 1 pL there is only a fraction of a molecule. Clearly, successive 1000-fold miniaturization leads to a situation where the test substance is undetectable, and hence this degree of miniaturization is impractical.

Sampling and Representative Samples. Biological samples are complex mixtures comprising cells and a diversity of large and small molecules (e.g., small metal ions, large macromolecular complexes). Most of the current clinical assays utilize samples in the 2–100 µL range. Successive reduction in the sample size to nL, pL, and fL volumes may lead to non-representative sampling errors, especially for constituents present at low concentrations or in low numbers (e.g., rare cell types). The extent of this problem has not been determined, but it is certain to limit the practical extent of miniaturization for detection of some types of analytes.

Cost-Effectiveness. Most microchips have been produced in glass or silicon for research applications in small experimental batches and so there is very limited information available on the eventual large scale manufacturing costs. Some concern has been voiced over the long-term economic viability of silicon, and there is now considerable interest in plastic as a alternative fabrication material (Chen and Chen 2000; Chen et al. 1999; Soper et al. 1999; Yu et al. 2000). Plastics could offer a degree of design and manufacturing flexibility not easily attained with silicon (e.g., low temperature bonding of components versus the high temperature anodic bonding used to cap silicon chips with glass covers). There would also the possibility of adapting the current range of low cost fabrication techniques available for plastics to the manufacture of microchips. One unknown may be the effect of the surface chemistry of plastics on the different bioanalytical reactions. For silicon chips this issue has already been encountered in PCR reactions (Shoffner et al. 1996) and it is not unreasonable to anticipate further surface chemistry problems with other fabrication materials.

The cost effectiveness of microchip devices in the future will greatly depend on the associated manufacturing costs. Very complicated designs involving immobilized reagents, on-board heating and cooling systems and expensive signal detectors will significantly diminish the ability to compete with conventional systems.

Integration. One of the key advantages of micro miniaturization is the ability to integrate sequential steps in an analytical process into a single device. This advantage has been verified by a number of different laboratories for a range of different analytical processes. Examples of microchip integration include sample preparation + PCR, cell lysis + PCR + electrophoretic sizing and capillary electrophoresis + electrochemical detection (Ekstrom et al. 2000; Mangru and Harrison 1998; Waters et al. 1998; Wilding P and Kricka 1999; Woolley et al. 1998). The resulting simplification of an analytical process will speed up analysis, and integration of the sample preparation will be particularly beneficial for genetic tests where contamination is a serious threat to the integrity of analytical results.

Point-of-Care versus Central Laboratory Applications. It is unclear if microchip-based analyzers will be used in analyzers intended for high throughput central laboratories or if they are better suited to point-of-care instrumentation. In the central laboratory anticipated benefits are smaller instrument footprints, hence reducing the space required. Also, based on the initial experience with the microchip-based CE analyzer (DNA LabChip™, Agilent Technologies, www.chem.agilent.com), it is expected that microchip instrumentation may be less costly and easier to operate. This will translate into greater flexibility in overall laboratory operations if staff can be quickly and easily trained to operate instrumentation.

Alternatively, the role of the microchip-based analyzer may be at the point-of-care. A continuing but slow trend in clinical testing is the devolution of testing from large centralized laboratories to the point-of-care (Price and Hicks 2000). The point-of-care includes the bedside, a clinic, a doctors' office or the home. The most successful tests in these settings have been blood glucose, pregnancy testing (based on testing for the hormone, human chorionic gonadotropin in urine) and the urine dipstick-type tests for multiple analytes (glucose, pH, urea, creatinine, nitrite, etc). The diversity of current microchip assays indicates that many highly complex tests, currently confined to the central laboratory, could be made available at the point-of-care. This could include genetic tests and this could be a controversial development. Irrespective of the technological feasibility of such analyzers, their introduction into routine use is determined by the prevailing regulation governing point-of-care testing.

4. Conclusions

Microchip technology has made rapid progress and the first examples of microchip-based analyzers for research applications have been commercialized. These types of device are poised to have a dramatic effect on the current practice of laboratory medicine, both in the central laboratory and in point-of-care settings. The widespread replacement of current analyzers by microchip analyzers will be dependent on the resolution of some remaining technological challenge, and confirmation of the economics of microscale testing. The potential scope of microchips in biological analysis is almost unlimited and will undoubtedly include most areas of laboratory medicine, biology and veterinary science. However, application of the technology will be limited by the ability to compete with non-microchip based devices in quality, cost and convenience.

References

Belgrader P, Benett W, Hadley D , Long G, Mariella Jr R, Milanovich F, Nasarabadi S, Nelson W, Richards J and Stratton P (1998) Clin Chem 44: 2191
Cheng J and Kricka LJ (eds) (2001) Biochip Technology, Gordon and Breach Scientific Publishers, Singapore

Cheng J, Shoffner MA, Hvichia GE, Kricka LJ and Wilding P (1996) Nucleic Acids Res 24: 380

Chen YH and Chen SH (2000) Electrophoresis 21: 165

Chen YH, Wang WC, Young KC, Chang TT and Chen SH (1999) Clin Chem 45: 1938

Chiem NH and Harrison DJ (1998) Clin Chem 44: 591

Colyer CL, Tang T, Chiem N and Harrison DJ (1997) Electrophoresis 18: 1733

Dolnik V, Liu S and Jovanovich S (2000) Electrophoresis 21: 41

Eggers M and Ehrlich D (1995) Hematol Pathol 9: 1

Ekstrom S, Onnerfjord P, Nilsson J, Bengtsson M, Laurell T and Marko-Varga G (2000) Anal Chem 72: 286

Gerry NP, Witowski NE, Day J, Hammer RP, Barany G and Barany F (1999) J Mol Biol 292: 251

Hofgartner WT, Huhmer AF, Landers JP and Kant JA (1999) Clin Chem 45: 2120

Kopp MU, Mello AJ and Manz A (1998) Science 280: 1046

Kricka LJ (1998) Clin Chem 44: 2008

Mangru SD and Harrison DJ (1998) Electrophoresis 19: 2301

Pease AC, Solas D, Sullivan EJ, Cronin MT, Holmes CP and Fodor SPA (1994) Proc Natl Acad (USA) Sci 91: 5022

Price CP and Hicks JM (eds) (2000) Point-of-Care Testing, AACC Press, Washington DC

Ross PL, Davis PA and Belgrader P (1998) Anal Chem 70: 2067

Shoffner MA, Cheng J, Hvichia GE, Kricka LJ and Wilding P (1996) Nucleic Acids Res 24: 375

Soper SA, Ford SM, Xu Y, Qi S, McWhorter S, Lassiter S, Patterson D and Bruch RC (1999) J Chromatogr A853: 107

Ueda M, Kiba Y, Abe H, Arai A, Nakanishiand H and Baba Y (2000) Electrophoresis 21: 176

Waters LC, Jacobson SC, Kroutchinina N, Khandurina J, Foote RS and Ramsey JM (1998) Anal Chem 70: 158

Waters LC, Jacobson SC, Kroutchinina N, Khandurina J, Foote RS and Ramsey JM (1998) Anal Chem 70: 5172

Wen J, Lin Y, Xiang F, Matson DW, Udseth HR and Smith RD (2000) Electrophoresis 21: 191

Wilding P and Ciaverelli C (2000) Hand-held sensors In: Price CP and Hicks JM (eds) Point-of-Care Testing, AACC Press, Washington DC, pp 41

Wilding P, Kricka LJ, Cheng J, Hvichia G, Shoffner MA and Fortina P (1998) Anal Biochem 57: 95

Wilding P and Kricka LJ (1999) Tibtech 17: 465

Woolley AT, Lao K, Glazer AN and Mathies RA (1998) Anal Chem 70: 684

Xue Q, Foret F, Dunayevskiy YM, Zavracky PM, McGruer NE and Karger BL (1997) Anal Chem 69: 426

Yu C, Svec F and Frechet JM (2000) Electrophoresis 21: 120

DNA Microarray Analysis of Gene Expression Profiles in Hepatocellular Carcinoma

Li Y., Dai J. L., Xie Y., Mao Y. M., Qiu M. Y., Cao X. Q., Fu W., Ying K., Xu H. and Huang S. D.

Abstract

Differential gene expression profiles between normal tissue and Hepatocellular Carcinoma using DNA microarray are described. DNA microarrays were prepared by spotting PCR products of 4096 human genes onto specially treated glass slides. The cDNA probes were prepared by labeling normal tissue mRNA and cancer tissue mRNA with Cy3-dUTP and Cy5-dUTP separately through reverse transcription. The arrays were then hybridized against the cDNA probe mixture and the fluorescent signals were scanned. The data obtained from repeated experiments were analyzed. Around 1000 genes exhibit differentially expression profiles in hepatocellular carcinomas from several specimens, some of which have been proved to involve hepatocellular carcinogenesis. Northern blot was used to verify the array hybridization data. Some unreported genes are undergoing further research concerning their function. This technology provides a powerful method to elucidate tumor-specific gene expression profiles in human cancer.

1. Introduction

Understanding biological processes such as diseases initiation at the gene level is becomeing a hotspot in molecular biology. Genes with different expression levels in different stages of growth, disease, cell cycle, response to stimulation, etc., may answer why these different stages occur. It is very important to screen out these special genes from the great amount of human genome. Conventional methods of northern or southern blot can not meet this requirement, but the advanced technique of DNA microarrays can do so. It allows monitoring the expression of thousands or tens of thousands of genes simultaneously in one hybridization experiment. Employing this technique, detection of differentially expressed genes and the construction of gene differential expression profiles are greatly facilitated.

In 1995, the first paper on gene expression profiles using DNA microarray was published by some biologists in Stanford University (Schena et al. 1995). Since then the DNA microarray technique has been employed widely to investigate the function of genes (DeRisi et al. 1997; Schena et al. 1996; Sharon et al. 1996; Wodicka et al. 1997), especially those genes involved in tumor generation and growth. There are already differential expression profiles of some kinds of tumors described in published papers, such as acute leukemia (Golub et al. 1999), melanoma (DeRisi et al. 1996), lung squamous cell carcinoma (Wang et al. 2000), Ewing's sarcoma (Welford et al. 1998), ovarian carcinomas (Wang et al. 1999), breast cancer (Osin et al. 1998; Sgroi et al. 1999), renal cell carcinoma (Moch et al. 1999), prostate cancer (Bubendorf et al. 1999), alveolar rhabdomyosarcoma (Khan et al. 1999; Khan et al. 1998). However, no such publication has mentioned hepatocellular carcinoma. There are many known or unknown genes relative to the pathological process of hepatocellular carcinoma. We tried to identify the genes involved in hepatocellular carcinogenesis for purposes of diagnosis, therapy and prevention.

2. Materials and Methods

2.1 Construction of microarrays

The microarrays consist of totally 4096 full-length or segmental novel and known genes provided by Shanghai BioDoor Gene Technology Company Limited. These genes were amplified through PCR reaction using universal primers and then purified (Schena et al. 1996). The PCR products were examined using agarose electrophoresis and dissolved in 3×SSC solution. These target genes were then spotted on silylated slides (TeleChem) by Cartesian 7500 Spotting Robotics (Cartesian). After spotting, the slides were hydrated (2 hr), dried (0.5 hr, RT), UV crosslinked (65 mJ/cm) and then treated with 0.2% SDS (10 min), H_2O (10 min) and 0.2% $NaBH_4$ (10 min). The slides were dried again and ready for use.

2.2 Probe Preparation

The tissue specimens concerned in this essay were provided by Changzheng Hospital, the Second Military Medical University, with the approval of the authority.

The method of total RNA extraction was modified from the original single step extraction of Piotr Chomczynski (Chomczynski and Sacchi 1987): the cancerous liver tissue and normal liver tissue stored in liquid N_2 were ground completely into tiny granules in a 100 mm ceramic mortar (RNase free) and homogenized in Solution D + 1% mercaptoethanol; after centrifugation the supernatant was extracted with equal volume of phenol : chloroform (1:1) twice and equal volume of acidic phenol : chloroform (5:1) once, discarding the organic phase each time; the aqueous phase was then precipitated by an equal volume of isopropanol at low temperature; the pellet was centrifuged and dissolved with Milli-Q H_2O.

The mRNAs were purified using Oligotex mRNA Midi Kit (Qiagen). The fluorescent cDNA probes were prepared through reverse transcription and then purified, referring to the protocol of M. Schena (Sharon et al. 1996; Stark 1991). The probes from normal tissues were labeled with Cy3-dUTP, those from cancerous tissues with Cy5-dUTP. Then the probes were mixed and precipitated by ethanol and resolved in 20 µL hybridization solution (5×SSC+0.2%SDS).

2.3 Hybridization and washing

After denaturing at 95 °C for 5 minutes, the probe mixture was added on the slides and covered with cover glass. The chips were incubated at 60 °C for 15-17 hours. The slide was subsequently washed in solutions of 2×SSC + 0.2% SDS, 0.1×SSC + 0.2% SDS and 0.1×SSC for 10 minutes each, then dried at room temperature.

2.4 Detection and Analysis

The chip was scanned by a ScanArray 3000 laser scanner GSI Lumonics at two wavelengths. The acquired image was analyzed by ImaGene 3.0 software (BioDiscovery). The intensities of each spot at the two wavelengths represent the quantity of Cy3-dUTP and Cy5-dUTP, respectively. Each ratio of Cy3 to Cy5 was computed. The two overall intensities were normalized by a coefficient according to the ratios of the located 40 housekeeping genes. We defined two standards to screen out each differentially expressed gene: 1. The absolute value of the natural logarithm of the ratio is greater than 0.69. 2. One or both of the raw intensity values of Cy3 and Cy5 are larger than 1000.

3. Results and discussions

3.1 Pretreatment

There are totally 4096 spots on one chip, including 1500 known genes and 2548 un-reported ones. In order to monitor the preparation and hybridization of the microarrays, we arranged some negative control spots in the arrays: rice U2 RNA gene (8 spots), HCV coat protein gene (8 spots), spotting solution (without DNA, 32 spots). We noticed that these negative control spots showed low intensity of signal after the hybridization, which proves the reliability of the statistics.

In order to test and reduce error occurrence in the statistics, normal hepatocyte mRNAs were labeled separately with Cy3-dUTP and Cy5-dUTP as mixed probes. After four hybridization repeats, we found that on average 3% of the spots showed obvious differences greater than 2 fold between the Cy3 and Cy5 signals and 1% had greater than 2.5 fold differences (Fig.1A), which is concordant with some reported statistics (Wang et al. 1999). This means that if we use 2.0 fold as the base line of the differential expression ratio, there will be about 30-100 false positive spots in one array. In further analysis we noticed that these spots appear randomly. These false positive spots were overcome by repeat experiments. The probability of a spot showing false positive in two experiments is reduced to 3%×3%=0.09%. So repeating the experiment greatly decreases the chance of false positive results. Moreover, using probes from different samples of the same kind of tissue decreases individual variations.

3.2 Genes with Differential Expression Between Normal Hepatocytes and Hepatocellular Carcinomas

Two individual primary hepatocellular carcinoma samples were chosen. Four groups of data were obtained by repeating each sample. There are 1210 genes which the ratios between the Cy3 signal and the Cy5 signal is greater than 2.0 while 903 genes show average ratios greater than 3.0 repeated in more than 3 groups. Among these 903 genes there are 109 genes showing higher expression in the cancerous tissue and 794 genes showing lower expression in the cancerous tissue. 379 off these 903 genes are known genes registered in Gene Bank (Fig.1B, Fig. 2A). Scatter plots with normal tissue show concentrated pattern (Fig. 1A), which means that the ratios are around 1; Scatter plots with cancer tissue shows a very wide distribution pattern (Fig. 1B), which means genes express differentlly in cancer cells compared with the normal cells.

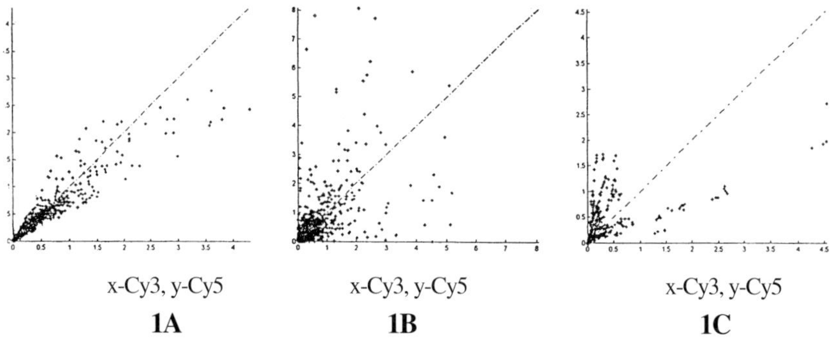

Fig. 1A. Scatter plots for normal liver (Cy5) / normal liver (Cy3) control hybridization,
1B. Scatter plots for hepatocellular carcinoma (Cy5) / normal liver (Cy3) hybridization (4096 elements),
1C. Scatter plots for hepatocellular carcinoma (Cy5) / normal liver (Cy3) hybridization (128x4 elements)

In an attempt to determine the reliability of the screened genes, another 512 array was designed which was composed of 92 genes from the 1210 differentially expressed genes 32 housekeeping genes and 4 negative control genes, each gene repeated in 4 spots. The mixed probes from normal liver and cancerous liver were used to hybridize against the array. The average ratio was obtained from four spots for which the CV value was usually less than 10% (Table 1). There were 85 genes whose average ratios between Cy3 and Cy5 are greater than 2.0, suggesting that the data is repeatable (Fig. 1C, Fig. 2B). Northern blot was also used to verify the array hybridization data (data not show).

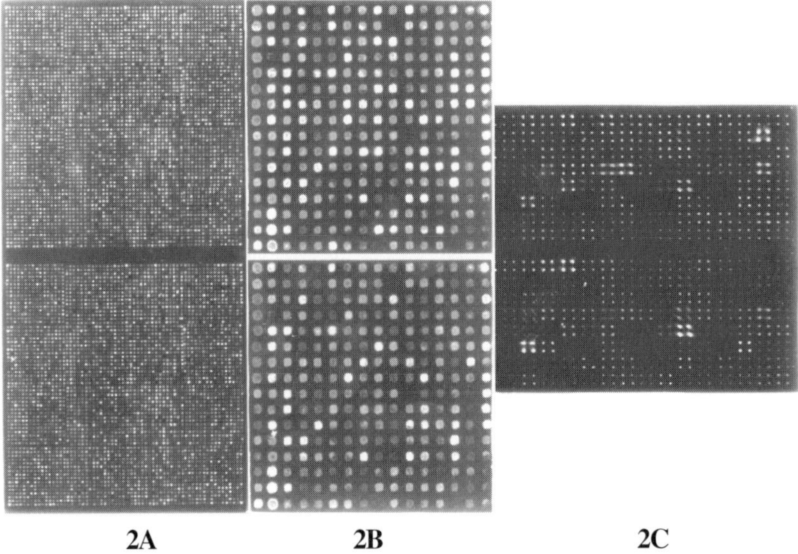

2A   2B   2C

Fig. 2A. Two-color fluorescent scan of human cDNA microarray containing 4096 elements. The array is hybridized with cDNA probe mixture of hepatocellular carcinoma (Cy5, up) and normal liver (Cy3, down) **2B.** Partial magnification of 1A **2C.** Two-color fluorescent scan of human a cDNA microarray containing 512 elements. The array is hybridized with cDNA probe mixture of hepatocellular carcinoma (Cy5, up) and normal liver (Cy3, down)

3.2.1 The Known Genes that are Over-Expressed in Cancer Cells

The old genes that are over-expressed in hepatocellular carcinomas include

(1) mitochondria related genes, such as Genebank ID, M22865, M22632, M22877, M22760, L15189, D13900, U59309, M37104, AF044956, AF070652 et al.
(2) peroxisome related genes, AF023462, U23942, X04076, X04096, M62401, S69189.
(3) glutathione S-transferase (GST) gene: J03817.
(4) fibronectin gene: X02761
(5) zinc finger protein gene: AF062347.
(6) metabolization related enzyme genes (omit)
(7) others

Some of these genes are discussed below.

(1) Peroxisome related proteins: Free radical injury on intracellular macromolecules can promote hepatocellular carcinogenesis (Perera et al. 1987). Peroxisome proliferators (PPs) are a class of non-genotoxic rodent hepatocarcinogens. PPs cause increases in the number and size of peroxisomes in liver and lead to enzyme induction, stimulate S-phase and suppress apoptosis by activating the PP-activated

receptor alpha in rodent liver (Hendrich and Piotot 1987; James et al. 1998). Our result shows that all of the peroxisome-related protein including P450 are over-expressed in hepatocellular carcinomas, which confirms the relationship between PPs and hepatocellular carcinoma.

(2) Enzymes of glutathione (GSH) metabolism: GSH is an antioxidant, radioprotector and radical scavenger. It also induces the formation of free radicals and oxidative damage to detoxify xenobiotics and carcinogens. Gamma-glutamyl transferase (GGT) and glutathione S-transferase (GST) are the key enzymes in GSH metabolism. They act as biochemical markers during hepatocellular carcinogenesis. Their expressions are enhanced shortly after their exposure to carcinogens and they play a role in multistage hepatocellular carcinogenesis. There is a strong presumpion that GGT-dependent oxidative damage may be similar to that of PPs as a promoter of hepatocarcinogenesis (Holden et al. 1998; Stark 1991). Our result supports this hypothesis.

(3) Fibronectin: Kemperman reported that fibronectin relates to the process of carcinoma cells forming metastases to hepatocytes. Adhesion between fibronectin receptors on metastasizing tumor cells and fibronectin on hepatocytes may lead to the formation of liver metastases (Kemperman et al. 1995). Our data shows that fibronectin is over-expressed in hepatocellular carcinomas. It may cause adhesion and metastases of liver tumor cells to other parts of the body by the same means.

Table 1. Partial results from 128x4 array. The average ratio is calculated from 4 repeat spots

Accession	Name	Average ratio	cv
M58603	nuclear factor kappa-B DNA binding subunit	0.17	5.15%
M31724	phosphotyrosyl-protein phoshatase (PTP-1B)	0.30	4.94%
AF131808	clone 24889	0.31	3.70%
X73460	rib protein L3A	0.34	1.61%
X56932	23KD highly basic protein	0.38	7.92%
AF050126	hyoxia-inducible factor 1 alpha subunit (HIF1A)	0.42	3.69%
Z37986	phenylalkylamine binding protein	2.85	13.96%
M37104	mitochondrial ATPase coupling factor 6 subunit	4.41	7.51%
U27460	uridine diphosphoglucose pyrophoshorylase	5.42	2.47%
X73424	propionyl-CoA carboxylase a subunit	5.50	2.82%
J03817	glutathione transferase M1B (GST1)	6.13	2.72%
Y12711	putative progesterone bindng protein	6.91	2.70%
AJ000388	calpain-like protease CANPX	7.59	4.53%
G37432	transferrin receptor	9.09	5.20%
X95384	translational inhibitor protein p14.5	10.04	3.75%
K03000	aldehyde dehydrogenase 1	10.30	3.91%
M22865	cytochrome b5	10.54	3.66%
M11718	alpha-2 type V collagen gene	14.26	2.06%

3.2.2 The Known Genes that Are Under-Expressed in Cancer Cells

The old genes that are down-expressed in cancer cells include
(1) genes encoding ribosome protein and translation-related perotein: X73460
    (L3), M77234 (S3), Z11531 (elongation factor-1-gamma).
(2) cytoskeleton protein coding genes:
    vimentin: Z19554, M25246, X56134.
    propomyosin: NM_003290.
    actin: L10373, M86406, M10277, M10278.
    tublin: K00558.
(3) zinc finger protein coding genes: D89859, X78925.
(4) protein tyrosine phosphatase (PTPase) coding genes: U16996, L38929,
    M33689, X93920, M93425.
(5) protein tyrosine kinase (PTKase) coding genes: NM003331, U07563, L40636.
(6) GDP-dissociation inhibitor of the rhoA GTP-binding protein (Ly-GDI) coding
    gene: L20688
(7) others
Some of the genes are discussed below.
(1) Protein tyrosine kinase (PTKase) and protein tyrosine phosphatase (PTPase):
PTKase and PTPase are involved in energy metabolism, cell proliferation and
stimulation of MHC class I molecule pathways. Holden mentioned that carcinoma
is associated with an inhibition of PTPase activity, which reduces the expression
of MHC class I antigen on the cell surface thereby allowing malignant cells to es-
cape immune surveillance (Hasmall et al. 1999). On the other hand, Tamura reported that PTEN, a tumor suppressor gene, is homology to PTPase and regulates
a complex signal transduction pathway. It inhibits cell migration and invasion by
directly dephosphorylating two key tyrosine-phosphorylated proteins. It also
maintains cell sensitivity to apoptosis. Loss of this gene prevents cells from
apoptosis (Tamura et al. 1999). Therefore, down-regulated expression of PTKase
and PTPase may accelerate the process of carcinogenesis.
(2) GDP-dissociation inhibitor of the rhoA GTP-binding protein (Ly-GDI): Small
GTP-binding proteins of the rho family are involved in the regulation of a variety
of cellular processes such as the organization of the cytoskeletal network and cell
signaling proliferation. Rho GTPase, particularly rhoA, are over-expressed in different types of human tumors. Fritz suggests that rho GTPase is involved in human carcinogenesis (Fritz et al. 1999). GDI inhibits rhoA activity by inhibiting
GDP/GTP exchange on rhoA (Scherle et al. 1993). Down-regulated expression of
Ly-GDI in hepatocellular carcinomas may lead to super-activity of rhoA, therefore
promoting carcinogenesis.

Hepatocellular carcinogenesis is a complex process for which the expression
profile differs greatly from normal hepatocyte expression. The number of over-
expressed genes is much less than that of down-expressed genes in hepatocellular
carcinomas, suggesting that the cell functions simplify in the cancerous cells.
Down-expression of ribosomal protein genes is another evidence. Wang reported

that the 149 mitochondrial transcripts studied are all over-expressed and 94 ribosomal protein genes all show a much lower expression level in ovarian tumor (Wang et al. 1999). This is consistent with our results in hepatocellular carcinoma.

3.2.3 Un-reported Genes that are Differentially Expressed in Normal and Cancer Cells

These new genes will be subjected to further experimentation to analyse their function.

Many differentially expressed genes were found using the DNA microarray technique, which can benefit further understanding of the molecular mechanism of hepaotic carcinogenesis and help discover more hepatocellular carcinoma markers to improve diagnoses and therapy as well. The DNA microarray technique will also be used in the large scale analysis of the primary functions of new genes.

References

Bubendorf L, Kolmer M, Kononen J, Koivisto P, Mousses S, Chen Y, Mahlamaki E, Schraml E, Moch H, Willi N, Elkahloun AG, Pretlow TG, Gasser TC, Mihatsch MJ, Sauter G and Kallioniemi OP (1999) J Natl Cancer Inst 91 :1758.

Chomczynski P and Sacchi N (1987) Biochemistry 162: 156-159.

DeRisi J, Penland, Brown LPO, Bittner ML, Meltzer PS, Ray M, Chen Y, Su YA and Trent JM (1996) Nat Genet 14: 457.

DeRisi JL, Iyer VR and Brown PO (1997) Science 270: 680.

Fritz G, Just I and Kaina B (1999) Int J Cancer 81: 682.

Golub TR, Slonim DK, Tamayo P, Huard C, Gaasenbeek M, Mesirov JP, Coller H, Loh ML, Downing JR, Caligiuri MA, Bloomfield CD and Lander ES (1999) Science 286 : 531.

Hasmall SC and Roberts RA (1999) Parmacol Ther 82: 630.

Hendrich S and Piotot HC (1987) Cancer Metastasis Rev 6: 155.

Holden RJ, Pakula IS and Mooney PA (1998) Med Hypotheses 54: 39.

James NH, Gill JH, Brindle R, Woodyatt NJ, Macdonald N, Rolfe M, Hasmall SC, Tugwood JD, Holden PR and Roberts RA (1998) Toxicol Lett 91:102-103.

Kemperman H, Driessens MH, Riviere G La, Meijne AM and Roos E (1995) Cancer Surv 24: 67.

Khan J, Bittner ML, Saal LH, Teichmann U, Azorsa DO, Gooden GC, Pavan WJ, Trent JM and Meltzer PS (1999) Proc Nat Amer. Sci 96: 13264.

Khan J, Simon R, Bittner ML, Chen Y, Leighton SB, Pohida T, Smith PD, Jiang Y, Gooden GC, Trent JM and Meltzer PS (1998) Cancer Res 58: 5009.

Moch H, Schraml P, Bubendorfl L, Mirlacher M, Kononen J, Gasser T, Mihatsch MJ, Kallioniemi OP and Sauter G (1999) Am J Pathol 154: 981.

Osin P, Shipley J, Lu YJ, Crook T and Gusterson BA (1998) Recent Results Cancer Res 152: 35.

Perera MI, Betschart JM, Virji MA, Katyal SL and Shinozuka H (1987) Toxicol Pathol 15: 51

Schena M, Shalon D, Dais RW and Brown PO (1995) Science 270: 467

Schena M, Shalon D, Heller R, Chai A, Brown PO and Davis RW (1996) Proc Natl Acad Science (USA) 93: 10614

Scherle P, Behrens T and Staudt LM (1993) Proc. Natl. Acad (USA) Science 90: 7568

Sgroi DC, Teng S, Robinson G, LeVangie R, Hudson JJ and Elkahloun AG (1999) Cancer Res 59: 5656

Sharon D, Mith SSJ and Brown PO (1996) Genome Research 6: 39

Stark AA (1991) Mutagenesis 6: 241

Tamura M, Gu J, Tran H and Yamada KM (1999) J Natl Cancer Inst 91:1820.

Wang K, Gan L, Jeffery E, Gayle M, Gown AM, Skelly M, Nelson PS, Ng WV, Schummer M and Hood L (1999) J Mulligan Gene 229: 101

Wang T, Hopkins D, Schmidt C, Silva S, Houghton R, Takita H, Repasky E and Reed SG (2000) Oncogene 19: 1519-28.

Welford SM, Gregg J, Chen E, Garrison D, Sorensen PH, Denny CT and Nelson SF (1998) Nucleic Acid Research 26: 3059

Wodicka L, Dong H, Mittmann M, Ho M-H and Lockhardt DJ (1997) Nature Biotechnology 14: 1675

DNA Hybridization Detection by Electrochemical Impedance Spectroscopy and Photoelectrochemistry

Li Q.-W., Luo G.-A, Feng J., Cai D. W. and Ouyang Q.

Abstract

Owing to the special properties of ssDNA and dsDNA molecules in their structures and electronic behaviors, they may give us many ideas for the fabrication of gene sensors and DNA-chips. In this work, photoelectrochemistry was first employed to characterize the behaviors of the self-assembled ssDNA probe modified electrode and the resultant dsDNA modified electrode. The obvious decrease in the photocurrent of the dsDNA modified electrode at open potential or a bias voltage indicated that photoelectrochemistry is another useful method for DNA hybridization detection. Concerning the special design of ssDNA probes, there is a further discussion on the relationship between the properties of DNA molecules and their photoelectric behaviors. In addition, the electrochemical impedance method was also employed to characterize and verify the hybridization event.

1. Introduction

The rapid progress in the Human Genome Project and people's strong desire for health nowadays has stimulated development of DNA-chips. Until now there have been about 3 billion functional gene segments for human body. With these genes assembled on a few DNA-chips it is expected that we can quickly check our health condition and predict diseases in the near future. However, some effort is still needed for the DNA-chip to enter into our real life. How to assemble as many gene probes as possible on a small chip with a size like a nail and how to simultaneously and rapidly recognize these specific binding events are the main problems for the fabrication of DNA-chips. Recently some techniques have been successfully applied to develop genosensors based on the combination of a suitable transducer, such as an electrochemical (Millan and Mikkelsen 1993; Millan et al. 1994; Palecek et al. 1998; Wang et al. 1997), chemiluminescent (Nguyen and Heffelfinger 1995), quartz crystal microbalance (Hashimoto et al. 1994; Okahata et al. 1992; Steel et al. 1998), fiber optical (Piunno et al. 1995), evanescent wave (Watts et al. 1995) or acoustic wave device (Su et al. 1996)with an immobilized ssDNA probe for the specific-sequence detection by DNA hybridization. With further understanding of the native physical and chemical properties of DNA, this will present us with some other new ideas for DNA hybridization detection.

Since Barton's group (Hall et al. 1996)first reported their exciting discovery that a long distance photo-induced electron transfer (> 4 nm) may be mediated by the DNA helix, whether the π-stack DNA system functions as a molecular wire or as an insulator has attracted much interest and dispute (Akin et al. 1996; Kelley and Barton 1998; Lewis et al. 1997). At least it has been generally accepted that this π-stack system should have some unique electric properties different from other kinds of biological macromolecules like proteins or carbohydrates. As we known electrochemical techniques can directly provide some information about the electron transfer properties of investigated molecules on the one hand. On the other hand, electrochemical methods can also be useful tools to qualitatively or quantitatively describe the interaction occurring between two molecules on the electrode interface according to the differences in their interfacial behaviors. As a result, some electrochemical techniques have been employed in this art. For instance, the voltammetric detection method (Palecek et al. 1998; Wang et al. 1997), which is based on the difference in the redox signal of a probe which merely interacts with the double DNA helix instead of single-strand DNA. Another method is the ac impedance method (Bardea et al. 1999), which is based upon the fact that electrode interfacial behaviors are different prior to and after the hybridization event since the charge distribution over an ssDNA molecule differs from that over its resulting dsDNA. Photoelectrochemistry is also an effective method to sense the change of an electrode interfacial state and electron transfer behavior of the molecules over the electrode surface (Imahori et al. 1998; Zhang et al. 1998). So in this chapter, we aim to study the photoelectric behaviors of ssDNA and dsDNA molecules and thus discuss the possibility to detect the DNA hybridization with this method.

2. Experimental

2.1 Reagents

Two 15-mer thiolated oligonucleotide probes were synthesized by the ShangHai Biochemical Engineering Corporation in China. One sequence only containing G and T was 5'-HS-TTTTTTTTTTGGGTTGGGTTGGGTT-3' its complementary oligomer had the sequence 3'-CCCAACCCAACCCAA-5'. The other only containing A and C was 5'-HS-CCCCCCCCCCAAACCAAACCAAACC-3' its complementary part was 3'-TTTGGTTTGGTTTGG-5'. The non-complementary DNA oligomers used was 3'TCTACGTCACAACTA. All the solutions used in the experiment were prepared with sterile distilled water. Electrochemical measurements were conducted with PBS buffer solution (NaCl 136.7 mmol L^{-1}, KCl 2.7 mmol L^{-1}, Na_2HPO_4 $12H_2O$ 9.7 mmol L^{-1} and KH_2PO_4 7.3 mmol L^{-1}). Hybridization reactions were performed in 2×SSC buffer.

2.2 Apparatus

Cyclic voltammetry and AC impedance experiments were carried out with the CHI660A Electrochemical Workstation (CH Instruments, Cordova, TN). The photoelectric measurement system operated with a Xe light source (Muller, light intensity is 121.4 mWcm^{-2}). The CHI660A was used to impose a bias voltage and record photocurrent simutaneously.

2.3 Electrode Preparation

A gold electrode with area of 0.3 mm^2 was first polished with ultrafine alumina, then put in a warm piranha solution (70% concentrated sulfuric acid and 30% peroxide solution) for 15 min and then rinsed with water thoroughly. The ssDNA modified electrode was prepared by immersing a clean gold electrode in a 20 µM solution of a probe oligonucleiotide in PBS overnight, then rinsing with water carefully before use. Hybridization was performed at 37°C for 1 hour in 2×SSC buffer. The concentration of the complementary target and the noncomplementary targent was 20 µM. Upon removal from the hybridization reaction solution the electrode was rinsed thoroughly with water and PBS solution. This was the so-called dsDNA modified electrode.

2.4 Procedures

Cyclic voltammetric and AC impedance experiments were performed in a single-compartment cell with a 10 mL volume. The working electrodes were prepared as described above. A saturated calomel electrode (SCE) and platinum wire served as reference and counter electrode, respectively. PBS solution was used as the supporting electrolyte. The photoelectrochemical measurements were carried out in a quartz crystal rectangular cell with a 5 mL volume. The other experimental conditions were the same as above.

3. Results and Discussion

Immobilization of DNA probe on a transducer surface with controlled quantity, coverage and orientation is crucial to the performance of genosensors. A variety of methods seems available to choose from, such as direct physical or chemical adsorption, cross-linking, encapsulation and covalent attachment, etc. Self- assembly of oligonucleic probes thiolated at 3' or 5' ends onto gold surfaces for hybridization and other purposes has recently become a popular method in this field. In our work, two kinds of specially devised probes were thiolated at the 5' end. Behind the –SH group there were ten T or C bases successively bonded in order to avoid coiling of the probe molecule over the electrode surface. As ssDNA and dsDNA are polyanions, when DNA molecules were self-assembled on gold electrode, owing to repulsion and space-inhibition between DNA molecules they can not be densely organized in a way similar to some thiol maptan molecules. The modified electrode surface could be characterized with the change of voltammetric and ac impedance behavior of a redox couple such as Fe^{3+}/Fe^{2+}. Figure 1 shows the voltammagrams of the gold electrode before and after modification with a monolayer of ssDNA and dsDNA. It could be observed that the redox behavior of Fe^{3+}/Fe^{2+} became less reversible with self-assembly of ssDNA and this situation became worse after the hybridization event occurred. It suggests that modification of ssDNA and further the formation of dsDNA on gold blocked the electron transfer between the redox couple and the gold electrode by electrostatic interaction.

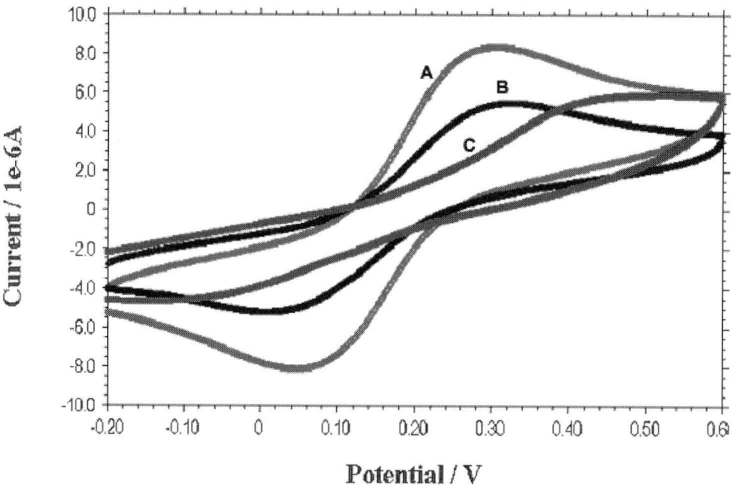

Fig. 1. Cyclic voltammetric curves of bare gold electrode (**A**) and modified electrode with ssDNA (**B**) and dsDNA (**C**)

Figure 2 shows the electrochemical impedance spectra of gold electrodes before and after modification with ssDNA and dsDNA molecules. It could be observed that the formation of self-assembly monolayers of ssDNA probe on gold and later the formation of dsDNA through hybridization with its complementary part resulted in an obvious increase in the diameters of the semicircles, indicating that the electrochemical reaction resistance for this redox couple was increased, which is in accord with the results shown in Fig. 1. Hybridization of both probes with their noncomplementary parts almost kept the original behaviors of their ssDNA. Figure 3 shows the effect of the hybridization time on the ac impendance behavior of the electrode and it further supplies evidence to show that the interfacial state of the electrode varies with the hybridization event.

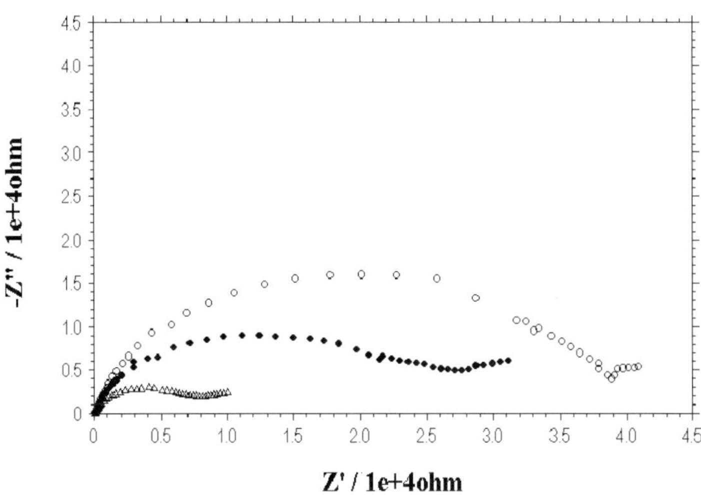

Fig. 2. Nyquist plots for Faraday impedance measurements on a gold electrode in the presence of 2 mM $[Fe(CN)_6]^{3-/4-}$: a) bare electrode, b) Au electrode with self-assembly of a monolayer of ssDNA (probe 2), c) dsDNA-Au electrode after hybridization with its complementary part

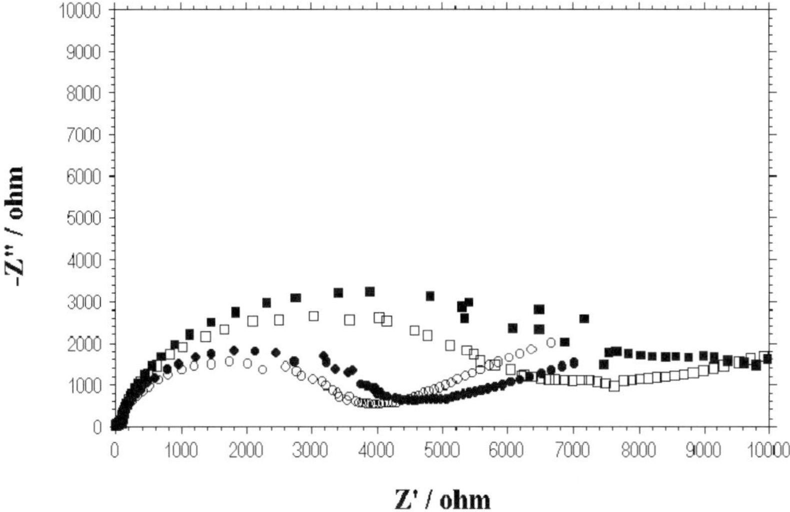

Fig. 3. The dependence of the interfacial behaviors of the electrode on hybridization time, open circle — ssDNA electrode, solid circle — hybridization time for 1min, open square — 5min, solid square — 30min

Figure 4 shows typical photoelectric response curves of the probe1 modified electrode in PBS buffer. It reveals that the photocurrent of the bare electrode in buffer was much lower than that of ssDNA modified electrode, while for the ssDNA modified electrode the photocurrent increased with the ion strength of the supporting electrolyte. The obtained cathodic photocurrent indicates electron transfer from the DNA molecule to the substrate electrode. It increased linearly with bias voltage up to 500 mV. As a higher voltage would damage the DNA probe by oxidation of the Au-S bond, all the photoelectric experiments were done below this critical value. Figures 5a and 5b show the photoelectric response behaviors of the probe 1 and probe 2 modified electrodes and their complementary dsDNA modified electrodes when they were illuminated with Xe light at 110 mW/cm^2 at 200 mV bias voltage. It revealed that the photocurrent of the dsDNA modified electrode decreased after hybridization. The decrease in amplitude increased with increasing bias voltage. Figure 6 is the action spectrum of the ssDNA and the dsDNA modified electrodes, respectively. Both electrodes reached their maximum photocurrents at about 400nnm, which was almost in agreement with the result obtained from plasma DNA (Zhang et al. 1998). Thus it confirmed that the observed photoelectric behaviors reflect the properties of immobilized DNA molecules.

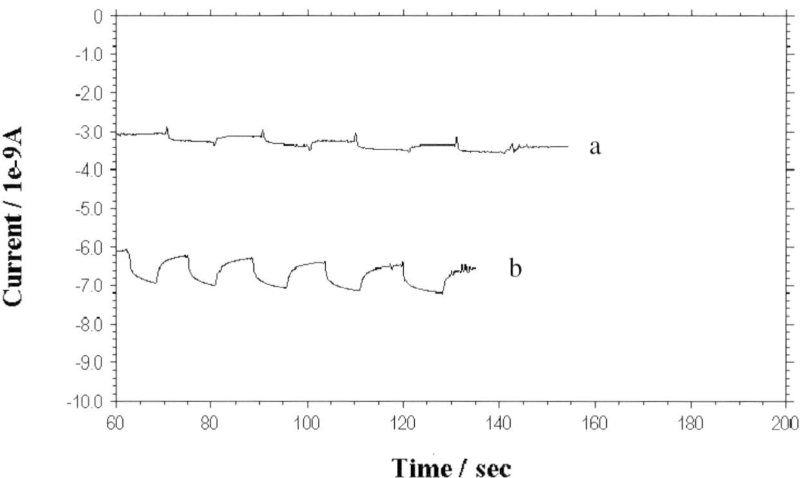

Fig. 4. Photoelectric currents of the bare gold electrode (**4a**) and the ssDNA modified electrode (**4b**) illuminated with a pulsed Xe light without any bias voltage

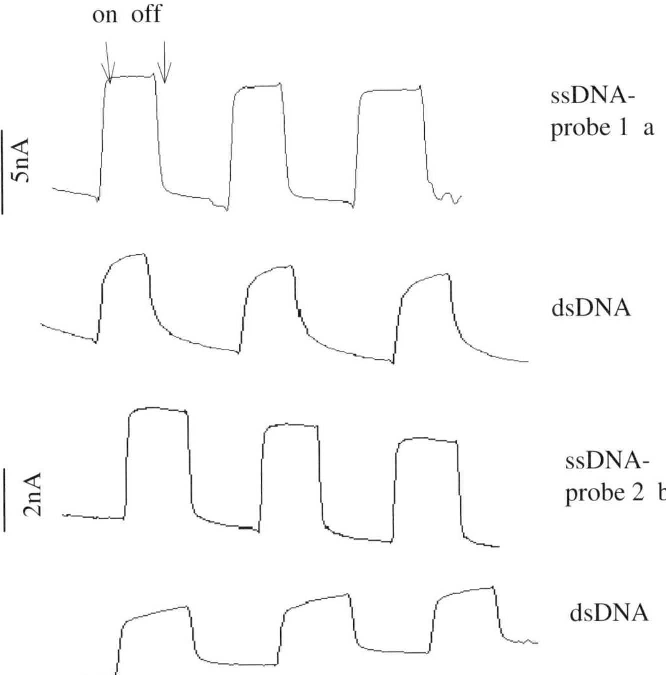

Fig. 5. Photoelectric behaviors of ssDNA and dsDNA modified electrodes

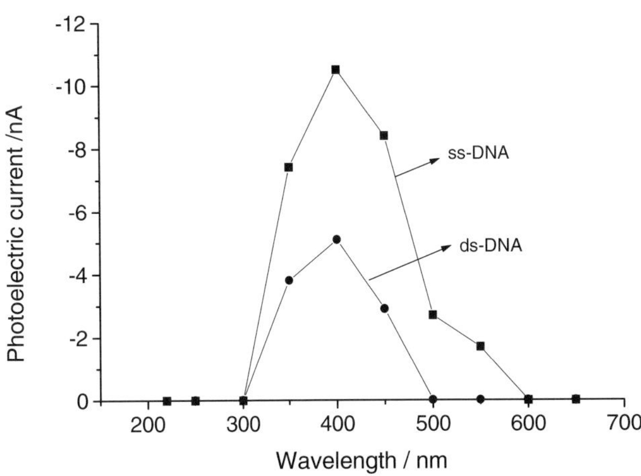

Fig. 6. The effect of light wavelength on the photocurrent of ssDNA and dsDNA modified electrodes

Self-assembly of a thiolated DNA probe on a gold electrode has now become a widely used method (Hashimoto et al. 1994; Okahata et al. 1992; Steel et al. 1998). When the ssDNA probe modified electrode is illuminated with a light source the observed response may either come from the absorption layer itself or from the whole behavior over the electrode interface. The DNA molecule can be considered as a semiconductor. It has been discovered that electron transfer along the DNA helix has a close relationship with the properties of G (Dandliker et al. 1998; Hall et al. 1996; Lewis et al. 1997; Saito et al. 1995). As the oxidation potential of G is lower than A and the other two bases which cannot be oxidized near the oxidation potential range of G and A, the DNA molecule with a sequence like probe 1 ought to be a good mediator for electron transfer. Owing to the lower oxidation potential of G, probe 1 might also be able to act as a photoactive molecule in which some G serves as donor. Under these conditions, when the probe 1 modified electrode was illuminated its photoelectric behavior was much large than the bare substrate electrode. When it hybridized with its complementary part the photocurrent of the dsDNA modified electrode decreased. This might be ascribed to the activity of G decreasing when specifically binding with C. This assumption seems in good agreement with the observed results. In order to further confirm the accuracy of this assumption, another probe was specially devised to have only A and T in its structure and employed to repeat the above experiment. To our surprise, we also observed that when the probe 2 modified electrode was hybridized with its complementary part, the photoelectric current of the dsDNA electrode decreased as well. This seemed unreasonable according to the above hypothesis. As there was much G in the complementary part, the addition of G in the dsDNA helix should be expected to result in an increase in the photocurrent of the dsDNA electrode. All the evidence indicates that the observed photoelectric phenomenon did not merely come from the DNA molecules but reflected the total interfacial properties of the modified electrodes. The DNA molecule is a kind of polyanion. The self-assembly of negatively-charged ssDNA and specifically formed double-stranded DNA molecules would alter the double-layer potential at the electrode surface, and further the Fermi energy difference at the electrode/solution interface, and eventually lead to different photoelectric behaviors for each kind of electrode. The changes in the interfacial state of the electrode before and after the hybridization event has been verified with the ac impedance measurement results shown in Fig.2.

In conclusion, owing to the special properties of the π-stack DNA system and large differences in the photoelectric behaviors of ssDNA modified electrodes and dsDNA electrodes, photoelectrochemistry could also be used as a powerful tool to rapidly detect DNA hybridization.

Acknowledgement

This project is financially supported by the National Natural Science Foundation of China.

References

Akin MR et al. (1996) Science 273: 475
Bardea A, Patolsky F and Willner I (1999) Chem Commun 21: 21-22
Dandliker PJ, Nunez ME and Barton JK (1998) Biochemistry 37: 6491
Hall DB, Holmlin RE and Barton JK (1996) Nature 382: 731
Hashimoto K, Ito K and Ishimori Y (1994) Anal Chem 66: 3830
Imahori H, Norieda H and Ozawa S (1998) Langmuir 14: 5335
Kelley SO and Barton JK (1998) Science 283: 375
Lewis FD, Wu TF and Zhang YF (1997) Science 277: 673
Millan KM and Mikkelsen SR (1993) Anal Chem 65: 2317
Millan KM, Sarraullo A and Mikkelsen SR (1994) Anal Chem 66: 2943
Nguyen Q and Heffelfinger DM (1995) Anal.Biochem 226: 59
Okahata Y, Matsunobo Y and Ijiro K (1992) J Am Chem Soc 114: 8299
Palecek E, Fojta M, Tomschik M and Wang J (1998) Biosensor & Bioelectrics 13: 621
Piunno PAE, Krull UJ and Hudson RHE (1995) Anal Chem 67: 2635
Saito I, Takayama M, Sugiyama H and Nakatani K (1995) J Am Chem Soc 117: 6460
Steel AB, Herne TM and Tarlov MJ (1998) Anal Chem 70: 4670
Su H, Chong S and Thompson M (1996) Langmuir 12: 2247
Wang J, Rivas G and Cai X (1997) Anal Chim Acta 347: 1
Watts HJ, Yeung D and Parkes H (1995) Anal Chem 67: 4283
Wu DG, Huang CH and Gan LB (1998) Langmuir 14: 3783
Zhang CY, Feng J and Ci YX (1998) Bioelectro & Bioenerg 46(1): 145

Data Mining: An Integrated Approach for Drug Discovery

Shi L. M. and Tong W. D.

Abstract

Drug discovery is an interdisciplinary endeavor that has greatly benefited from advances in chemistry, pharmacology, microbiology, biology and biochemistry. The process of bringing a molecule to the marketplace is very complex, time-consuming and expensive. To remain competitive in the industry, the pharmaceutical company is forced to adopt new technologies to shorten the time required for drug discovery and development. Recent advances in genomics, combinatorial chemistry, high throughput screening and biochip technology are having a revolutionary impact on various stages of the drug discovery process. The synergy among these fields is emerging. However, a common challenge to researchers in each of these fields is how to turn the massive raw data that have been accumulating at an explosive rate into useful information and knowledge in order to more efficiently guide the process of drug discovery. This chapter addresses the needs and commonalities of data mining in these areas. The principles of these technologies are briefly described at first, followed by a discussion of the rational integration of data mining tools into pharmaceutical research. Advances in the analysis of massive gene chip data are also discussed.

1. Introduction

Drug discovery is an interdisciplinary endeavor that has greatly benefited from advances in chemistry, pharmacology, microbiology, biology and biochemistry. The process of bringing a molecule to the marketplace is very complex, time-consuming and expensive (Drews 2000; Smith 1992). According to information provided at the web site (www.pharma.org) of the Pharmaceutical Research and Manufacturers Association (PhRMA), on average it costs a company $500 million and 15 years to get a new drug from the laboratory to patients. As shown in Fig. 1, among the 5,000~10,000 compounds that are synthesized and screened, only about 250 enter preclinical animal testing and five of them advance to human testing, i.e., the Phase I - Phase III clinical trials. Only one of those five compounds is approved by the Food and Drug Administration (FDA) for marketing. Furthermore, only three of the ten approved drugs can generate profit that is enough to overcome the research and development costs.

In the discovery phase scientists use different technologies to search for novel compounds capable of curing diseases. Historically this searching process was based on trial and error and serendipitous discovery. However, it has become more and more rational and systematic through the use of sophisticated technologies such as combinatorial chemistry (CombiChem) (Gallop et al. 1994; Gordon et al. 1994), high throughput screening (HTS) (Cox et al. 2000), DNA microarray (Schena 1999; Shi 1999; Southern et al. 1999) and computer-aided drug design (Charifson 1997; Hansch et al. 1995).

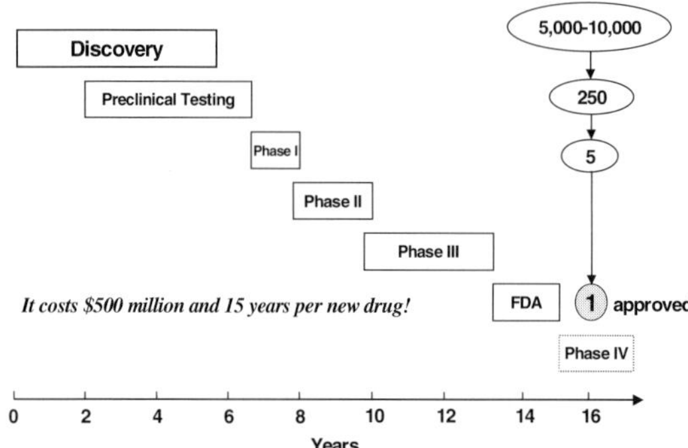

Fig. 1. The drug discovery and development process is time-consuming and expensive

To ensure the safety and efficacy of the approved drugs the pharmaceutical companies and the FDA are taking extraordinary measures to conduct and regulate the clinical trials for drug candidates after enough animal testing data have been collected and evaluated. The FDA requires an IND (Investigational New Drug) application from the pharmaceutical company and clinical trials cannot be initiated until the IND is approved.

In Phase I the drug candidate is evaluated for safety in healthy volunteers in a population of 20~80 people per trial. A trial is conducted with a single dose of the drug, beginning with small doses. If the compound is shown to be safe, multiple doses of the drug are evaluated in other clinical trials with escalating dosages. If the compound is shown to be safe in Phase I, it is advanced to Phase II. The main purpose of Phase II is to determine the efficacy of the compound in a population of 100~300 patient volunteers, although safety is also studied. Data are collected and analyzed to determine whether the compound is efficacious and safe for the intended patient population. If so, the drug is advanced further to Phase III clinical trials. Under these trials, safety and efficacy are further evaluated in a much larger population of as many as 1,000~5,000 patient volunteers, including the elderly, patients with multiple diseases, those who take other drugs and patients whose organs are impaired.

If the drug candidate shows superior efficacy and safety profile compared to available medicine already approved by the FDA for the same diseases, the pharmaceutical company can file an NDA (New Drug Application) with the FDA for approval of marketing this drug.

In some cases drugs may be approved for sale before the FDA is totally satisfied with the safety data that have been collected. However, the pharmaceutical company is required to continue to conduct additional safety studies and track the safety of the drug for the treated patients. This process is called Phase IV, or post-market surveillance.

The pharmaceutical company is under enormous pressure to enrich its "pipe-line" with novel drug molecules so as to maintain a competitive edge in the marketplace. On average the pharmaceutical industry invests 15~20% of its annual revenue in research. Competition in the pharmaceutical industry forces the company to adopt innovative technologies that are able to bring a molecule to the marketplace in a shorter period of time.

Figure 2 illustrates the impact of four cutting-edge technologies on drug discovery and the demand for data mining efforts to utilize the huge amount of data being generated by these technologies.

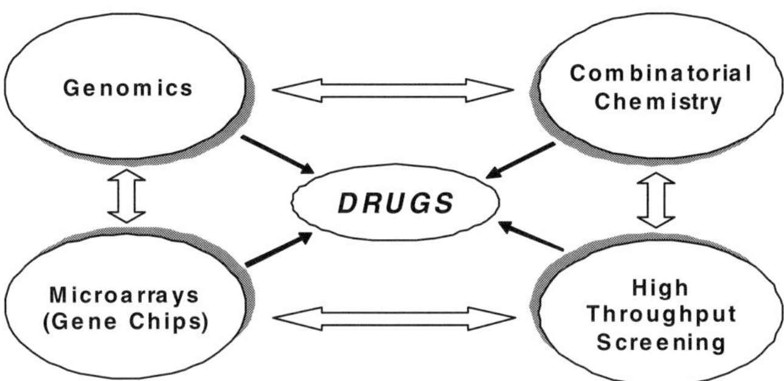

Fig. 2. Integrating data mining in the drug discovery process. Novel drug discovery technologies are generating a huge amount of data that needs to be mined.

Genomics (Firestein 2000) studies genes and their functions on a large, or genome-wide, scale. With the dramatic progress of the Human Genome Project and other genome projects researchers are faced with a huge collection of gene sequence data. The storage, annotation, retrieval and analysis of the sequence data have been the major tasks of the new field of bioinformatics (Searls 2000). Sequence similarity analysis provides clues to gene functions based on the assumption that genes with similar sequence should have similar functions. Homology modeling has proven useful in providing useful protein 3D structure information for drug discovery from the crystal or NMR structure of a related or homologous protein. Once the active site of the protein is identified, structure-based molecular modeling tools can be used to design ligands that can bind effectively to the active site. By either inhibiting the undesired biological function of a (mutated) protein or bringing on the desired biological cascade, protein-ligand binding is a major mechanism by which most drugs work.

Combinatorial chemistry synthesis technology (Gallop et al. 1994; Gordon et al. 1994) is able to provide a large number of candidate compounds for lead identification and optimization in a much shorter period of time. However, even with combinatorial synthesis technology, chemists will still never be able to synthesize all theoretically possible compounds. Therefore there is a great need of designing diverse combinatorial libraries rationally using computational chemistry algorithms (Dean and Lewis 1999).

The compound libraries based on combinatorial synthesis or corporate historical collections are assayed in an array of HTS systems (Cox et al. 2000) to identify hits - compounds that elicit a positive response in a particular assay. Because of the large number of compounds available for screening and the number of assays that have been developed for different therapeutic purposes, researchers are overwhelmed with a huge amount of data about the compounds and their activities. Data mining technology (Frawley et al. 1992) is being used to help researchers to understand the screened data and identify the complicated relationships between chemical structure and biological activity (Kenakin and Lutz 1999; Shi et al. 1998, Shi et al. 2000, a Shi et al. 1998, Weinstein et al. 1997).

Another dramatic technological breakthrough in biomedical research is the DNA microarray or gene chip (Schena 1999; Shi 1999; The Chipping Forecast, Nature Genetics. January supplement issue (1999) 21: 1-60), a subset of the more broadly defined biochip technology. With the help of DNA microarrays the mRNA expression levels of thousands of genes can be monitored simultaneously, as opposed to traditional single gene monitoring. The parallel nature of the DNA microarray technology enables researchers to investigate the biological effects of drug candidates on a genome-wide scale. Genes whose expression levels are dramatically altered after drug treatment may be the drug target, or genes related to the drug target, although more study is needed for target validation. Making sense of the huge amount of gene expression data from DNA microarray technology has become one of the most challenging tasks of today's bioinformaticians (Searls 2000).

2. Experimental Details and Results

Data Mining and Knowledge Discovery in Databases. As we described earlier, the modern drug discovery and development process is generating a huge amount of data through the utilization of new technologies in genomics, combinatorial chemistry, high throughput screening and DNA microarray or biochip. A common challenge facing researchers in each of these fields is how to turn the massive raw data that has been accumulating at an explosive rate into useful information and knowledge in order to guide the process of drug discovery. It is becoming clear that the rate-limiting stage is no longer the generation, but the interpretation, of data.

Data mining and knowledge discovery in databases (KDD) are new terms that have been used to describe the research efforts of turning raw data into useful knowledge for decision-making. Usually these terms are used interchangeably, although, generally speaking, data mining focuses on the algorithms and KDD deals with the whole process that includes data storage, retrieval, pre-processing and analysis. Data mining or KDD is defined as "the nontrivial extraction of implicit, previously unknown and potentially useful information from data" (Frawley et al. 1992). "Nontrivial" means that data mining is not a simple task. To get the hidden, previously unknown information from data requires special expertise.

Figure 3 illustrates the data mining and KDD process. The first step is to collect data relevant to the problem and maintain it in a database. In many situations the raw data is the repository of corporate historical records, and therefore, data mining researchers have no control of how and when the data should have been collected. However, in some other cases researchers can design the experiments a priori based on sound statistics in order to collect data that is most appropriate for data mining. For each situation, raw data needs to be cleaned (e.g., eliminating obvious outliers and incomplete records) and pre-processed (e.g., data normalization). The core part of the KDD process lies in the step of knowledge discovery - it is at this step that useful knowledge is discovered by using various algorithms such as machine learning, statistical and visualization techniques. The results of data mining should be presented in a form that can be easily understood by humans. After certain post-processing the results can be integrated to support decision-making.

The following is a partial list of statistical, machine learning and visualization techniques that we have been using in data mining for drug discovery (Shi et al. 1998).

- Principal Component Analysis and Regression (PCA and PCR)
- Multidimensional Scaling (MDS)
- Pattern Recognition (PR)
- Hierarchical Cluster Analysis (HCA)
- Non-Hierarchical Cluster Analysis (K-Means, Jarvis-Patrick)
- Correlation Analysis (Pearson, Spearman)
- Multiple Linear Regression (MLR)
- Partial Least-Squares Regression (PLS)
- Soft Independent Modeling of Class Analogy (SIMCA)
- K-Nearest Neighbors (KNN)
- Artificial Neural Networks - Back Propagation and Self-Organizing Maps
- Multivariate Adaptive Regression Splines (MARS)

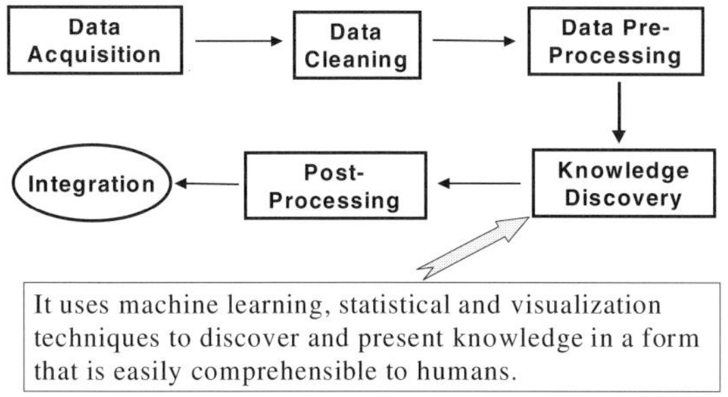

Fig. 3. Data mining and KDD process

- Classification and Regression Trees (CART)
- Genetic Algorithms (GA)
- Cross-Validation (CV) and Bootstrapping
- Clustered Image Maps

Our experience is that there is no universal "best" method that is suitable for all problems. Instead, each method has its own advantages and disadvantages. For a particular problem one method may be better than the other. It is up to the researcher to identify the most appropriate method(s) for her/his particular problem, usually by exploring all or most of the methods with the data.

Representing Molecules in the Computer for Comparison. As shown in Fig. 1, it is usually required to screen 5,000~10,000 compounds before one drug can be put into the marketplace. Can we increase our chance of success with the use of new technologies including data mining techniques? The answer is yes. The general idea is to make the computer compare molecules and then to identify and eliminate those molecules unlikely to become a drug at a very early stage of the drug discovery process. In doing so, research efforts can be focused on the drug-like molecules. However, computers only understand numbers. The first thing we need to do is to describe molecules in terms of numbers instead of the 2D or 3D chemical structures (images) that human beings are accustomed to. Depending on what kind of molecular features we are interested in or are relevant to drug discovery, a molecule can be described in many different ways. Figure 4 shows my view of the relationship between various methods of characterizing molecules. In short, there are three levels of molecular representation methods, i.e., the S-P-A.

The first level of molecular description is the structural descriptors (S). Most of these descriptors are molecular fragment-based. For example, in MDL Information Systems Inc.'s 2D MACCS key approach, a molecule is described by an array of 1's and 0's that describe the presence or absence of a set of predefined substructural fragments, respectively. These fragments, such as phenol or other functional groups, are selected based on the analysis of large chemical databases so that they can be used to effectively enhance the performance of chemical database searching. The similarity or distance between two molecules can be easily calculated based on the two integer arrays by using a distance metric, e.g., the Euclidean distance measurement. 3D screens, on the other hand, are pharmacophore-based. They describe the critical 3D distances between pairs of pharmacophore features, including hydrogen-bond donors and acceptors, hydrophobic centers and aromatic centers, which are considered essential for a molecule to elicit certain type of biological response. CoMFA (Comparative Molecular Field Analysis) fields [4] are also 3D molecular features, but they are based on the electrostatic and steric interactions between a molecule and a probe atom, usually an sp^3 carbon with one positive charge. Comparison of molecules can be easily performed on the similarity of the CoMFA fields, as will be discussed in more detail in the QSAR section.

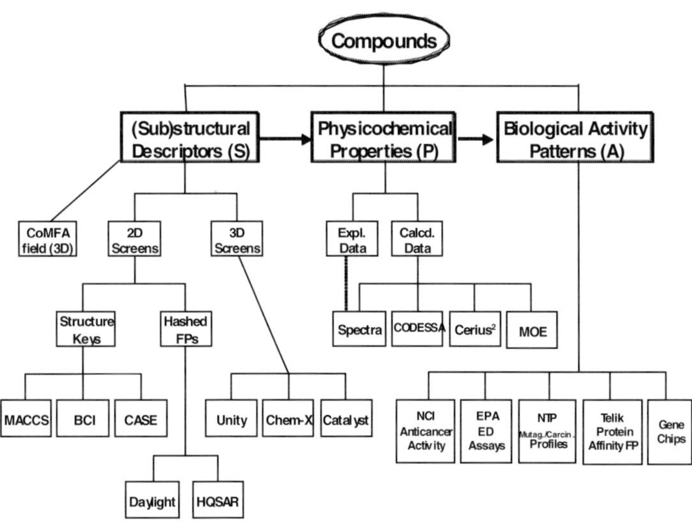

Fig. 4. S-P-A: three levels of molecular features, including structural descriptors (S), physicochemical properties (P) and biological activity patterns (A)

The second level of molecular description is the physicochemical properties (P) that can be either determined experimentally or calculated theoretically, e.g., various spectral data (IR, UV/Vis, NMR, and MS), molecular weight, water solubility and lipophilicity logP. In contrast to the S-level representation that is mainly based on the fragments of a molecule, generally speaking, the P-level molecular representation describes the whole molecular property. Many P-level properties can be readily calculated based on molecular structure.

The most complicated level of molecular representation method is the biological activity pattern (fingerprint or profile) (A) of a compound across a series of biological test systems instead of one single test system. The activity pattern can be considered as an overall picture of the biological effects of the tested compound. One well-known example is the anticancer activity patterns generated by the U.S. National Cancer Institute's anticancer drug discovery program (Shi et al. 2000a; Weinstein et al. 1997). In 1990 the NCI established a primary screen in which compounds are tested in vitro for their ability to inhibit growth of 60 human cancer cell lines from different organs, including leukemia, melanoma, breast cancer, colon cancer, etc. The 60 activity values (one for each cell line) make up the unique activity fingerprint for each compound. We have demonstrated (Shi et al. 1998a; Shi et al. 2000a; Shi et al. 1998b) that, although the anticancer activity against one specific cancer cell line does not provide much information, the activity patterns across the 60 cell lines as a whole encode rich information on the mechanism of action of screened compounds. Compounds with similar mechanisms of action tend to show similar activity profiles. The activity pattern can be used to predict the mechanism of action of new drug candidates. The U.S.

Environmental Protection Agency initiated a program for the identification of endocrine disruptors (Blair et al. 2000; EPA's Endocrine Disruptor Screening and Testing Advisory Committee (EDSTAC) (1998) Final Report, http://www.epa.gov/ scipoly/oscpendo/history/finalrpt.htm). Under the program, each compound will be tested in an array of more than 10 assay systems. The resulting activity profile would be a unique fingerprint about the endocrine disrupting ability of the compound. The U.S. National Toxicology Program (NTP) screens compounds for mutagenicity and carcinogenicity across different species. The results represent the toxicological (mutagenic and carcinogenic) profiles of tested compounds. Telik Inc. developed a unique method for predicting ligand binding to proteins by affinity fingerprinting from a small panel of reference proteins (Dixon and Villar 1998; Kauvar et al. 1995). After preliminary testing of over 300 proteins from a variety of sources, the authors selected panels of 8~18 proteins that displayed the broadest binding affinities for a set of over 5,000 compounds. MLR was used to build "computational surrogates" to predict the binding potencies of additional molecules. The affinity fingerprint database, which provides a rich source of data defining "operational similarities" among the proteins, can be used for efficient prescreening of a large number of compounds against target proteins in order to select promising candidates for further study. The most recent biological activity pattern is the DNA microarray or gene chip data. Each molecule can be represented by changes in the mRNA expression levels of thousands of genes before and after treatment by the compound. Gene Logic Inc. (www.genelogic.com) is building such a database called Gene Express. Many pharmaceutical companies are building their own gene expression databases.

The three levels (S-P-A) of molecular features are central to drug discovery. The A-level activity data, including desired therapeutic activity and undesired toxicological profile, contains the most relevant information for drug discovery and, in many cases, can be accurately predicted based on the S- and P-level properties through the methods of QSAR and related methodologies (Hansch et al. 1995; Shi et al. 1998a).

QSAR: Predicting Biological Activity from Chemical Structure. The main purpose of QSAR (quantitative structure-activity relationship) study (Hansch et al. 1995; Shi et al. 1998a). is to build a quantitative relationship, or model, between chemical structure information and biological activity. As shown in Figure 5, there are two main steps in the QSAR approach: training and prediction. In the training step molecular descriptors for a set of compounds with known biological activity are calculated and used as the independent variables (X) for correlation with biological activity, i.e., the dependent variable (Y). There are many statistical and artificial intelligence methods, including MLR, PCA, PCR, PLS, SIMCA, KNN, HCA, MDS, MARS, CART, ANN and GA, that can be used to derive the mathematical formula of the model, f. After the model is built it is expected to be applied for activity prediction of new compounds not included in the training set. The prediction is based on the calculated molecular descriptors and the mathematical model. Since molecular descriptors can be readily calculated for molecules that have not yet been synthesized, the QSAR prediction can be applied for virtual screening of drug candidates. Compounds with high predicted activity and

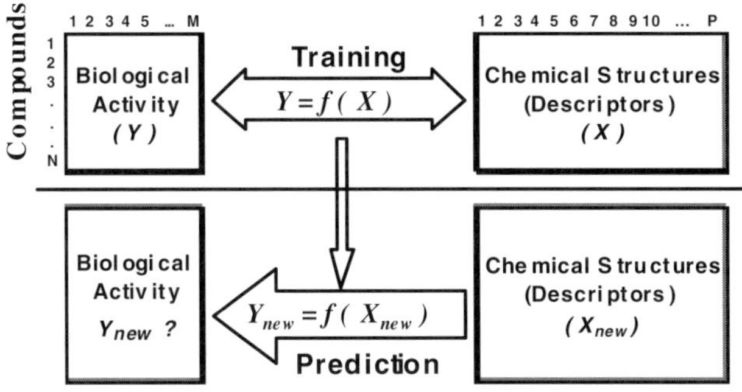

Fig. 5. Training and prediction processes in QSAR study

low predicted toxicity are more likely to show good profiles in biological testing and therefore should be given higher priority for synthesis and screening.

There are many types of QSAR methods depending on the different ways in which molecular descriptors are calculated and the mathematical methods used to derive the model. Among all available QSAR methods, CoMFA (Cramer III et al.1988) has been proven to be one of the most reliable and useful approaches when the mechanism of action of the molecules under study is based mainly on the non-covalent interaction between ligands and the receptor (a Shi et al. 2000). Figure 6 shows the general procedure for performing a CoMFA/QSAR analysis.

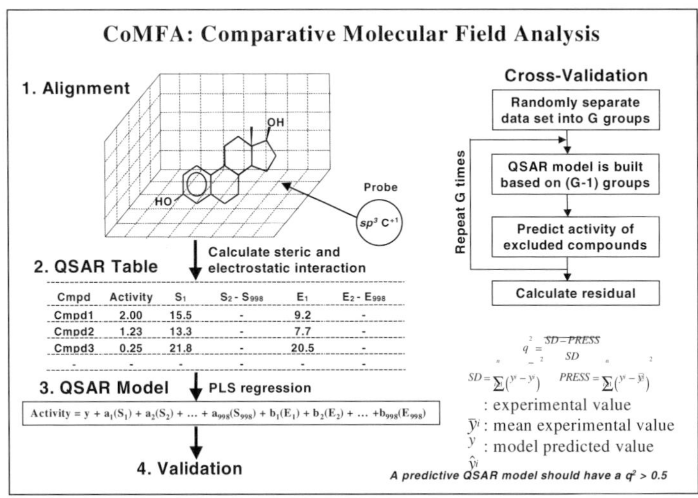

Fig. 6. General procedure for performing a CoMFA/QSAR analysis

First, partial atomic charges are calculated for all the molecules in the training set, which are then energy-minimized. The energy-minimized molecules are then aligned in a 3D space based on their similarity or commonly shared structural features. One compound, usually the lead compound, is used as the template and other molecules are aligned on top of the template, one-by-one. The alignment step is critical to the success of a CoMFA analysis. If molecules are not aligned in a meaningful way the comparison among them based on molecular fields (as discussed below) will not make sense. Alignment rules should be developed based on understanding of the structure-activity relationship (SAR) and the commonality of the training set compounds. These rules have to be followed for all the molecules in the training set and the prediction set.

Secondly, a probe atom, usually an sp^3 carbon with one positive charge, is used to sample the steric and electrostatic potential around each molecule in a predefined 3D grid box. The 3D box should cover all the molecules and the grid spacing is usually 2 Å. Therefore, each molecule can be represented by an array of hundreds to thousands of steric and electrostatic interaction energy terms. These energy terms are used as independent variables for QSAR correlation.

Thirdly, the partial least-squares regression (PLS) method is used to reduce the dimensionality of the descriptor space to a few principal components (PCs) and to correlate the reduced descriptors with biological activity. A QSAR model is just a function between the steric and electrostatic interaction energy and the biological activity.

Fourthly, the model built in Step 3 needs to be validated. Mathematically, it is possible to fit SAR of the training set very well when there are many descriptors, as is the case in CoMFA. However, fitting known experimental data is not the main purpose of QSAR. The predictive ability of a QSAR model is what makes QSAR useful for drug discovery. A standard approach for QSAR model validation is called cross-validation (CV). In CV, the training set is randomly separated into G groups of approximately equal size. One group of compounds is left out from the training process. A QSAR model is built based on the (G - 1) groups and used to predict the activity of the compounds left out in the training process. The whole model-building and prediction process is repeated for each of the G groups until each group is left out and predicted once and only once. The residual between model prediction and experimental value is calculated for each molecule and is summed to assess the model predictability. This CV procedure is mainly used to determine the optimal number of PCs for the PLS calculation. The number of PCs in the model that has the lowest prediction residual is used to build the final CoMFA/QSAR model that encompasses all the training set molecules. Usually, the CV is done by the leave-one-out (LOO) method, i.e., each molecule in the training set is considered a separate group during CV. q^2, as determined in the LOO-CV process, is the most widely accepted statistical parameter for measuring the predictive power of a QSAR model. Generally speaking, a predictive QSAR model should have a q^2 value of greater than 0.5.

A very useful feature of CoMFA/QSAR is that the modeling results can be visually examined through CoMFA contours. Basically, CoMFA contours summarize QSAR results in a graphical way so that they can easily tell researchers what kind of steric and electrostatic changes can be made to the molecule for improved activity. The QSAR results can be used to guide the next round of synthesis for activity optimization, e.g., through the design of focused combinatorial libraries. We have used this approach routinely in our discovery work.

Designing Combinatorial Libraries of Drug-like Molecules for HTS. The development of HTS technology (Cox et al. 2000; Kenakin and Lutz 1999) make it possible for researchers to screen thousands of samples each day in the search for hits. The pharmaceutical company is now facing the problem of providing the HTS system with a sufficient flow of compounds. In addition to the company's own inventory compounds, outside compound collections are being purchased. In fact, there are many companies that are specialized in providing compounds for HTS, with the number of available compounds ranging from a few thousands to about 1 million. These compounds are mostly historical collections from many academic institutions. Although the diversity of these compounds is generally good, price and sample quality are always of concern.

Traditional medicinal chemistry is based on the one-compound-in-one-experiment approach, i.e., each time only one substituent is changed and only one new analog is synthesized. Therefore productivity is low. Typically, a medicinal chemist can only synthesize a few dozens to a few hundred analogs per year. This output is definitely not enough for modern HTS technology. Combinatorial chemistry (Gallop et al. 1994; Gordon et al. 1994) promises to provide a large number of compounds based on the combinatorial synthesis strategy.

Figure 7 illustrates the general idea of combinatorial chemistry. For the benzodiazepine library we defined four positions where substitutions can be made to create new compounds. If there are 10 possible substituents at each of the four R-groups, then theoretically, there are $10 \times 10 \times 10 \times 10 = 10,000$ analogs that can be made. Combinatorial synthesis, also called parallel synthesis or automated synthesis, tries to make all possible combinations.

The earliest strategy of combinatorial synthesis was to make all analogs in a mixture and use the screening protocol to screen against the mixture. Because some smart deconvolution techniques were employed, the hits can be purified and identified after biological screening. Nowadays, combinatorial chemistry efforts in the pharmaceutical industry tend to focus on making pure individual analogs that can be screened individually, although in an HTS format. Early combinatorial chemistry efforts focused on synthesizing a huge number of combinations, usually hundreds of thousands or even millions of compounds in a mixture. However, the quantity of each compound was very limited. Structure identification and confirmation was one of the bottleneck problems. Currently, the industry tends to focus on much smaller libraries, usually consisting of a few hundred to at most a few thousand analogs per library.

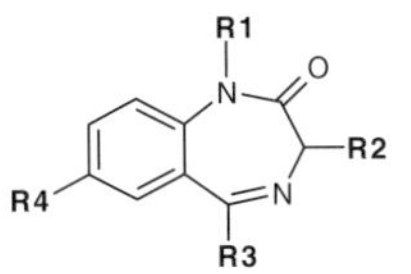

Fig. 7. The benzodiazepine library and the combinatorial synthesis strategy

In fact, there are many more possible substituents available at each of the four substitution positions shown in Fig. 7. Therefore, the theoretical number of analogs is much larger than 10,000. As we mentioned earlier, even with the help of ombinatorial chemistry, synthetical chemists are still unable to synthesize all possible combinations.

One basic task of library design is to select a subset of compounds from all possibilities so that the sub-library has a maximum diversity to represent the whole library and at the same time a subset that can be synthesized in a combinatorial way. This is not a trivial task since the simple "cherry-picking" approach will not work.

Library design includes several steps: the enumeration of all possible candidates, the calculation of structural descriptors for each molecule, the definition of the diversity or similarity metrics for compound comparison and the selection of a mathematical method to generate the sub-library. Software packages are available for the automated enumeration of all possible compounds at a very high speed. Then a set of descriptors, such as those shown in Fig. 4, is calculated for each molecule. Compounds can be compared based on the similarity of their descriptors. The dimensionality of the descriptor space is usually reduced to three for visual inspection. This is generally done with the help of principal component analysis (PCA), where the variance in many descriptors is captured by the first few PCs. Library design or diversity selection method needs to be constrained by the combinatorial nature of synthesis, i.e., if in one substitution position a substituent is present in one of the selected compounds then all the combinations of the subsituent with all other substituents at other positions should be considered. This constraint results in a selection of sub-optimal diversity but can be implemented in an automated synthesis format. This library design method is based on the diversity of products.

Another straightforward approach to library design is the reagent-based method. In this method a diverse subset of substituents is selected for each substitution position based on the "cherry-picking" diversity analysis of the individual reagent list for each subsitution position. Then the subsets of reagents (subsituents) are combined to generate the final library of desired size. Generally speaking, the reagent-based library design approach yields libraries with lower diversity compared to those designed by the product-based approach.

"Cherry-picking" diversity analysis (Dean and Lewis 1999) is useful in selecting a subset of the corporate library for assay purpose or in selecting vendor compound libraries to fill the "holes" of one's own inventory. The selection meets the requirement of maximum diversity but ignores the possibility of combinatorial synthesis in the laboratory. The three main approaches for library design and compound selection are cell-based, distance-based and cluster-based.

When designing combinatorial libraries or selecting vendor compounds for purchasing, researchers can make the library or selection biased to "drug-like" candidates. By comparing the distributions of certain physicochemical properties of a large set of drug and non-drug molecules that have advanced to at least Phase II clinical trials, researchers found some characteristic differences between the two types of molecules. The most famous finding was the "rule of 5" by Lipinski et al. (Lipinski et al. 1997), which states that poor absorption or permeation is more likely when the molecular weight is over 500, the lipophilicity (logP) is over 5, there are more than 5 H-bond donors (expressed as the sum of OH's and NH's) and there are more than 10 H-bond acceptors (expressed as the sum of O's and N's). (Compound classes that are substrates for biological transporters are exceptions to the rule.)

Compounds with one of these four features are flagged out in company databases and be given a lower priority in the discovery and development process. As a general guideline, Lipinski's "rule of 5" is a very useful indicator of drug-likeness for a candidate compound. During the process of combinatorial library design researchers can use property filters to eliminate compounds that do not have the desired properties possessed by most of the drugs. In other words, we can focus on the drug-like molecules to increase our chance of success.

Making Sense of Massive Gene Chip Data. The power of DNA microarray (or gene chip) lies in its parallel nature of monitoring thousands of genes simultaneously, which is in sharp contrast with traditional tools in biology. For more detailed discussions of the gene chip technology, see, e.g., other chapters in this book, a monograph edited by Schena (Schena 1999), a supplement issue of Nature Genetics (Southern et al. 1999) and a frequently visited and informative gene chips Web site (www.gene-chips.com) maintained by Shi (Shi 1999). Basically, a gene chip is a miniaturized bio-analytical system that is based on the nucleic acid base-pairing or hybridization principle, i.e., A-T and G-C pairing for DNA, and A-U and G-C pairing for RNA. DNA chips are fabricated by high-speed robotics, generally on glass but sometimes on nylon substrates. DNA probes with known identity are used to determine complementary binding of the target DNA samples to be analyzed, thus allowing massively parallel gene expression and gene discovery studies. There are two major application formats for the DNA chip technology, i.e., the identification of gene sequences and the quantification of gene expression levels. An experiment with a single DNA chip can provide researchers with information on thousands of genes simultaneously - a dramatic increase in throughput.

In terms of the property of arrayed DNA probes there are two major variants of the DNA chip technology. In format I probe cDNA (500~5,000 bases long) is immobilized to a solid surface such as glass using robot spotting and exposed to a set of targets either separately or in a mixture. In format II an array of oligonucleotide (20~25-mer oligos) or peptide nucleic acid (PNA) probes is synthesized either in situ (on-chip) or by conventional synthesis followed by on-chip immobilization. The array is exposed to labeled sample DNA, hybridized and the identity and/or abundance of complementary sequences are determined.

Gene chip technology is finding many applications in drug discovery. One application is to "individualized therapy" based on the genetic fingerprint of individual patients. Why do some drugs work better in some patients than in others? Why are some drugs even highly toxic to certain patients? The answer is the genetic difference between patients. Genetic polymorphisms in drug-metabolizing enzymes, transporters, receptors and other drug targets have been linked to interindividual differences in the efficacy and toxicity of many drugs. The goal of pharmacogenomics (Evans and Relling 1999) is to find correlations between therapeutic responses to drugs and the genetic profiles of patients, thereby enhancing drug discovery and providing a stronger scientific basis for optimizing drug therapy on the basis of each patient's genetic profile. Patients should only be given medications that have been demonstrated effective on the population of patients with similar genetic profiles.

Just a couple of years ago, gene chip technology was only available to a few laboratories. However, it has become more and more accessible to molecular biologists and will definitely evolve to a standard tool in any modern laboratory of molecular biology. Whether it is being used for gene sequence identification (e.g., single neicleotide polymorphism or SNP detection) or for gene expression level quantification (e.g., mRNA expression profiling), gene chip technology is generating a huge amount of raw data. It is safe to say that the bottleneck problem in the field is no longer the generation, but the interpretation, of this massive data.

The mRNA levels of thousands of genes are determined in each microarray experiment. The data from multiple experiments can be arranged into an N by P data matrix format, as shown in Fig. 8 , wherein N is the number of genes being monitored by the microarray and P is the number of experiments (or samples) that have been run through with the microarray system. Each cell of the data matrix represents the expression level of a particular gene in a particular experiment (sample). The value of the cell is usually represented in terms of fold difference (FD) compared to a control sample, where a value greater than 1 means over-expression and a value less than 1 means under-expression. Each column of the matrix represents the results of one experiment, i.e., expression levels of N genes, whereas each row represents the expression profile of that gene across the P experiments (samples). Conceivably, samples (experiments) can be compared based on the similarity of columns and genes can be compared according to the similarity of rows.

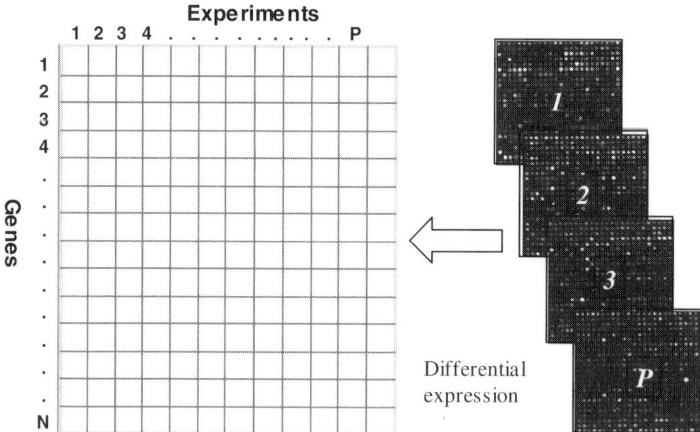

Fig. 8. Data format of a DNA microarray: Each gene is characterized by its expression profile across the P experiments (samples), whereas each experiment (sample) is characterized the mRNA expression levels of the N genes monitored by the microarray.

There are three typical questions that researchers may ask about the microarray data. The first question is which genes are differentially expressed. To answer this question we need to examine the expression data and sort out those genes that are either over-expressed or under-expressed. Because of the inherent variation in the microarray experiment, it is difficult to make a judgement solely based on the results of one experiment. How far away should the FD value be before we can say that a gene is differentially expressed? Statistically, it is necessary to perform replicates to assess the variation that comes from the microarray technology itself so that an experimental fluctuation will not be taken as a real difference in gene expression. In practice, some researchers use a cutoff FD value of ≥ 2.0 to determine which genes are differentially expressed.

The second question is which genes are co-expressed. Here, two genes are compared in terms of their expression profiles across the P experiments. In other words, two rows are compared. The Euclidean distance and the Pearson correlation coefficient are two widely used metrics for gene comparison. The assumption is that co-expressed genes should have similar expression profiles across experiments under different conditions.

The third question is which genes form gene clusters. At this level, data for multiple genes across multiple experimental conditions are considered simultaneously. In other words, the data matrix shown in Fig. 8 is taken into consideration altogether. Multivariate data mining methods that have been applied to analyze microarray data include PCA, HCA, MDS, self-organizing maps (SOM), support vector machines (SVM), singular value decomposition (SVD) and fuzzy logic. Among them, the HCA method is most widely used.

Cluster analysis groups objects in terms of their similarities. Hierarchical clustering produces a cluster tree in which similar objects are grouped together, whereas dissimilar objects are distant from each other in the hierarchy (Shi et al. 2000a; Shi et al. 1998b). The clustering results can also be visually inspected by rearranging the genes and experiments in their respective cluster order. This clustered color image representation of data was first used by Weinstein et al. (Weinstein et al. 1997) and has gained wide acceptance in the microarray community (Eisen et al. 1998). We have used this approach to analyze the anticancer activity data of 25,023 compounds against the 60 cancer cell lines (Shi et al. 1998a; Shi et al. 2000b).

PCA is a technique that takes linear combinations of the original variables (columns) of a data matrix such that the first PC explains as much of the overall variation as possible, the second PC explains the next most important variation subject to being orthogonal to the first, and so on (Shi et al. 2000b). PCA is often used as a data reduction technique to examine large databases. Scherf et al. (Kenakin and Lutz 1999) reported the microarray data for 1,375 genes across the 60 cancer cell lines used in the NCI anticancer drug discovery program and analyzed them with HCA. We did a PCA analysis on the same data set and the first three PCs are plotted in Fig. 9. The relative positions of the 60 cancer cell lines can be clearly identified. The major observations based on mRNA expression data show dramatic similarity with those from our previous clustering analysis of the 60 cell lines based on the "pharmacological" profiles – the anticancer activity patterns of 25,023 compounds across the 60 cell lines (Shi et al. 2000a). Our previous analysis showed that the six leukemia cell lines formed a cluster. So did the six melanoma cell lines. The seven colon cell lines also grouped together. For other types of cell lines it was hard to see clear grouping by organ of origin. Furthermore, two cell lines previously labeled as breast cancers showed a dramatic similarity to each other and clustered much closer to the melanoma cell lines than to any other cell lines. All these observations are also obvious from Fig. 9. The "coincidence" indicates that mRNA expression data is indeed useful for drug discovery since it reflects the pharmacological profiles of screened compounds. The two "breast" cancer cell lines, pointed out with a question mark in Fig. 9, are now labeled as of "unknown" origin by organ by the NCI drug anticancer drug discovery program. We also applied HCA and other methods to analyzing the same data set (data not shown).

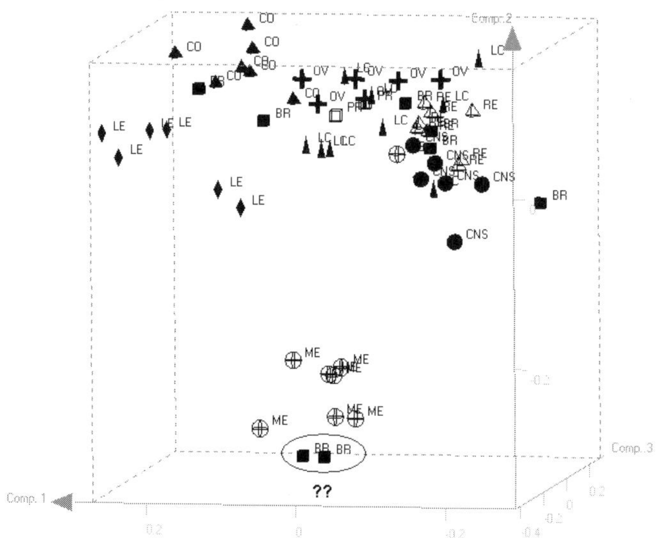

Fig. 9. Similarity of 60 cancer cell lines based on mRNA expression levels of 1,375 genes. Two cell lines, previously labeled as "breast" cancers, are more likely to be melanomas. BR: breast cancer; CNS: central nervous system cancer; CO: colon cancer; LC: non-small cell lung cancer; LE: leukemia; ME: melanoma; PR: prostate cancer; OV: ovary cancer; RE: renal (kidney) cancer.

3. Discussion

Advances in genomics, CombiChem, HTS and DNA microarray are having a revolutionary impact on the way pharmaceuticals are being discovered and developed. As an integrated component of drug discovery efforts, data mining or KDD technology plays a crucial role in manipulating the huge amount of data being generated and in extracting useful knowledge for decision-making. On the genomics side, the accumulation of gene sequence data combined with bioinformatic tools makes it possible to identify novel gene targets for drug discovery. Structural genomics and computational tools for protein structure prediction promise to provide accurate 3D protein structures of drug targets. Based on the protein 3D structure, computational chemists can rationally design molecules that fit the active site. Virtual combinatorial libraries of drug-like molecules can be generated and screened in silico with QSAR methods. Sub-libraries of candidates with high predicted activity can be made using combinatorial synthesis strategy.

The hypothetical drug target needs to be validated for its suitability for drug intervention. Reproducible and dose-dependent phenotypic changes in isolated cells elicited by a compound that modifies the target constitute the lowest level of target validation (EPA's Endocrine Disruptor Screening and Testing Advisory Committee (EDSTAC) (1998) Final Report, http://www.epa.gov/scipoly/ oscpendo/history/finalrpt.htm). If the phenotypic changes can be induced reproducibly in an animal model the degree of validation is higher. Of course, the best validation of the drug target would be the desired clinical efficacy of the compound that modifies the target. Development of HTS assays based on validated target models allows rapid identification of hits. It is also desirable to have HTS systems to identify compounds that have favorable ADMET (adsorption, distribution, metabolism, excretion and toxicity) profiles.

The rapid accumulation of HTS data makes it more and more important to relate chemical structure to biological data and to derive QSAR models for lead optimization. SAR from HTS data provides critical information to guide the next round of synthesis effort. Novel QSAR methods that are able to handle a large number of compounds are being developed. Virtual screening using QSAR models will continue to be used in drug discovery practices.

It has been reported that the similarity principle, which states that similar molecules should yield similar biological activity, does not always hold (Dixon and Villar 1998). Therefore it is important to design combinatorial libraries with not only structural diversity but also biological activity diversity.

Gene chips, as a powerful technology for gene sequence detection and expression profiling, provide a new tool for compound characterization. A drug candidate can be characterized by the changes it makes on the expression levels of thousands of genes. A reference database of such profiles, such as the one being built by Gene Logic Inc., can serve as a reference tool to tell good drug candidates from bad ones. Gene expression data will likely be used to predict the therapeutic outcome of drug candidates.

The emergence of protein microarrays, especially the recent work by MacBeath and Schreiber of Harvard University (MacBeath and Schreiber 2000), will add another powerful tool for molecular biologists to directly study gene functions at protein level on a genome-wide scale. The Harvard team printed more than 10,000 proteins on a glass slide using the robot that has been used for making DNA microarray. It has been demonstrated that the protein microarray can be used to identify protein-protein and protein-drug interaction. It will be extremely useful for identifying novel drug targets and possible side effects of drug candidates. The advantage of protein chip over DNA chip is obvious: after all, it is proteins, not DNA or RNA, that carry out the vast majority of chemical reactions in cells. I believe that it is a true breakthrough for proteomics and for drug discovery. Conceivably, the large amount of data being generated by protein chips also needs to be mined by the methods discussed in this chapter.

References

Blair RM, Fang H, Branham WS, Hass BS, Dial SL, Moland CL, Tong W, Shi L, Perkins R and Sheehan DM (2000) Toxicol Sci 54: 138-153

Charifson PS (ed.) (1997) Practical Application of Computer-Aided Drug Design, Marcel Dekker, New York

Cox B, Denyer JC, Binnie A, Donnelly MC, Evans B, Green DV, Lewis JA, Mander TH, Merritt AT, Valler MJ and Watson SP (2000) Prog Med Chem 37: 83-133

Cramer III RD, Patterson DE and Bunce JD (1988) J Amer Chem Soc 110: 59-5967

Dean PM and Lewis RA (eds) (1999) Molecular Diversity in Drug Design, Kluwer, Dordrecht

Dixon SL and Villar HO (1998) J Chem Inf Comput. Science 38: 1192-1203

Drews J (2000) Science 1960-1964

Eisen MB, Spellman PT, Brown PO and Botstein D (1998) Proc Natl Acad Sci (USA). 95: 14863-14868

EPA's Endocrine Disruptor Screening and Testing Advisory Committee (EDSTAC) (1998) Final Report, http://www.epa.gov/scipoly/oscpendo/history/finalrpt.htm

Evans WE and Relling MV (1999) Science 286: 487-491

Firestein S (2000) Nature 404: 552-553

Frawley WJ, Piatetsky-Shapiro G and Matheus CJ (1992) AI Magazine 13: 57-70.

Gallop MA, Barrett RW, Dower WJ, Fodor SP and Gordon EM (1994) J Med Chem 37: 1233-1251

Gordon EM, Barrett RW, Dower WJ, Fodor SP and Gallop MA (1994) J Med Chem 37: 1385-1401

Hansch C, Leo A and Hoekman D (1995) Exploring QSAR, Volume 1. Fundamentals and Applications in Chemistry and Biology, American Chemical Society, Washington DC

Kauvar LM, Higgins DL, Villar HO, Sportsman JR, Engqvist-Goldstein A, Bukar R, Bauer KE, Dilley H and Rocke DM (1995) Chem Biol 2: 107-118

Kenakin T and Lutz MW (1999) Quantitative Molecular Pharmacology and Informatics in Drug Discovery, Wiley, New York

Lipinski CA, Lombardo F, Dominy BW and Feeney PJ (1997) Adv Drug Delivery Rev 23: 3-25

MacBeath G and Schreiber SL (2000) Science 289: 1760-1763

Schena M (eds) (1999) DNA Microarrays: A Practical Approach, Oxford University Press, NewYork

Scherf U, Ross DT, Waltham M, Smith LH, Lee JK, Tanabe L, Kohn KW, Reinhold WC, Myers TG, Andrews DT, Scudiero DA, Eisen MB, Evans EA, Pommier Y, Botstein D, Brown PO and Weinstein JN (2000) Nature Genetics 24: 236-244

Searls DB (2000) Drug Discovery Today 5: 135-143

Shi LM, Fan Y, Myers TG, O'Connor PM, Paull KD, Friend SH and Weinstein JN (1998a) J Chem Inf Comput Sci 38: 189-199

Shi LM, Myers TG, Fan Y, O'Connor PM, Paull KD, Friend SH and Weinstein JN (1998b) Mol Pharmacol 53: 241-251

Shi LM (1999) The gene chips (DNA microarrays). Web site http://www.gene-chips.com, featured in Science 285: 799

Shi LM, Fang H, Tong W, Wu J, Perkins R, Blair R, Branham W and Sheehan D (2000a) J Chem Inf Comput Sci, submitted

Shi LM, Fan Y, Lee JK, Waltham M, Andrews DT, Scherf U, Paull KD and Weinstein JN (2000b) J Chem Inf Comput Sci 40: 367-379.
Smith CG (1992) The Process of New Drug Discovery and Development. CRC Press, Boca Raton
Southern E, Mir K, Shchepinov M, Nature Genetics. January supplement issue (1999) 21: 1-60.
Weinstein JN, Myers TG, O'Connor PM, Friend SH, Fornace Jr AJ, Kohn KW, Fojo T, Bates SE, Rubinstein LV, Anderson NL, Buolamwini JK, Van Osdol WW, Monks AP, Scudiero DA, Sausville EA, Zaharevitz DW, Bunow B, Viswanadhan VN, Johnson GS, Wittes RE and Paull KD (1997) Science 275: 343-349

Protein Array Detection with Nanoparticle Fluorescent Probes by Laser Confocal Scanning Fluorescence Detection

Sun B.-Q., Yi G.-Sh., Xing W.-L., Chen D.-P., Zhou Y.-X., Cheng J.

Abstract

The new and versatile protein microarray technology allows high-throughput screening for gene expression, drug screening and disease detection. An protein microarray using antibodies modified by ZnS-coated CdSe quantum dot (ZnS/CdSe QDs) nanoparticle and detected with a conventional laser confocal scanner is described here. Antibody targets labeled with fluorescent QDs rather than organic fluorophore probes have many desirable properties. The fluorescent signals of the sandwich immunoassay conjugate were detected by a laser confocal scanner. A diode laser was used to excite the fluorescent signals efficiently while bovine serum albumin was used to eliminating nonspecific binding sites. The specificity of the QDs-labeled immunoglobulin (IgG) was tested in an experiment using goat IgG and human IgG samples. The result was consistent with the binding specificity in a sandwich-type assay. The potential of this method to function as a simple and efficient readout strategy for microarray is discussed.

1. Introduction

Immunoassay technology is limited to the analysis of a few thousand assays per day, even if the microtiter plate format is used. However, this is about to change rapidly as we enter the microarray era, where the assay format allows tens of thousands of assays to be run in parallel (Ekins 1990). The first analytical breakthrough has come with the use of DNA chips, but an analysis of the proteome requires a different approach and the development of so-called protein chips (Carl 2000).

Protein chips will have to use a 'probe' (such as an antibody) on the surface of a silicon chip to be able to catch native and post-translationally modified proteins (such as antigens) which are usually labeled with fluorophores (Andrew et al. 2000; Macbeath and Schreiber 2000). Detection is again via laser excitation of the fluorophores and CCD-based optical readout.

The measured fluorescent signal typically results from the accumulated sum of all labels present in the target region, including contributions from both specific binding events and backgrounds from cuvettes, solvents, optics, etc. Scattering interfaces is aggregated by the small Stokes shifts (24-50 nm) of conventional fluorophores, so strong backgrounds usually have a dominant effect on the sensitivity of fluorescent immunoassay.

Fluorescent quantum dots (QDs) are attractive objects for scientific research on three-dimensionally confined systems. Several attractive fluorescent characteristics of QDs suggest that they might be of value as fluorescent labels and as a read-out method in the area of biomedical and clinical chemistry research. Quantum confinement effects in QDs lead to narrow emission spectra whose wavelength can be tuned over a wide range of particle size (Murray et al. 1993). For instance, the intrinsic spectral width of CdSe/ZnS QDs was reported to be about one third as wide as that of a fluorescent sphere, which minimized the potential overlap between the emission band and the excitation band and increased the detection sensitivity. Furthermore, the emission of QDs is generally 50-100 times more stable than that of most traditional organic dyes against photo-bleaching, which makes it possible to average the signal for extended time periods to lower the detection limit. In addition, the excitation band of QDs is rather wide and different colours are emitted in the same QD material with different dot sizes, which allows the use of single excitation source for multi-channel detection (Chan and Nie 1998; Marcel et al. 1998).

At present two different approaches have been reported for linking biomolecules and core-shell QDs with CdSe cores and CdS or ZnS shells. Marcel et al. coated the surface of QD with silica and then linked biomolecules to the surface (Marcel et al. 1998). Chan et al. directly linked proteins to QD surfaces using mercapto- acetic acid and showed that QDs bearing transferrin underwent receptor-mediated endocysis into HeLa cells (Chan and Nie 1998).

QDs are the most attractive luminescent materials because of their superior properties. However, their application in the biological area has just begun (Taylor et al. 2000). The object of this chapter is to explore the possible application of QD-labeled biomolecules to the immunoassay on glass protein chip model. A microminiaturized immunoassay on a glass slide was carried out using CdSe/ZnS QD labeled antibody with laser confocal scanning fluorescence detection. Immunoassay was carried out on a glass chip using the well-known sandwich assay. First, antibody was covalently bound to a glass chip. Second, the bound antibody was allowed to capture antigen specifically. Third, the detection antibody labeled with QDs was allowed to bind selectively to the captured antigen. Our results indicate that QDs are suitable fluorescent materials for immunomolecule labeling as substitutes for traditional organic dyes.

2. Instruments, Materials and Methods

A model LS-50B luminescence spectrometer (Perkin-Elmer, Norwalk, CT) was used for the fluorescence analysis of QDs and QD-immunomolecule conjugates. The readout of the immunoassay with labeled antibody on the glass slide and the calculation of the fluorescence intensity were performed on a GSI Scanner Array 4000 (GSI Lumonics) equipped with two lasers (532 nm, 635 nm).

CdSe/ZnS QDs were prepared according to the literature (Margaret and Philippe 1996). Human IgG, goat IgG, rabbit anti-human IgG and bovine serum albumin (BSA) were purchased from Jingke. D, L-mercaptosuccinic acid (MSA), 3-aminopropyltrimethoxysilane (APTES), diisopropylethylamine, gluotaraldehyde, 1-ethyl-3-(dimethylamino- propyl) carbodiimide (EDC) and xylene were acquired from Acros and used without additional purification. The water used was prepared by a Milli-Q system (Millipore, Bedford, MA).

Preparation of Capture Antibody Substrate Glass. Microscope chips were first soaked in chrome pickle for 2h, copiously rinsed with deionized water and dried in a stream of nitrogen. Then they were ultrasonically cleaned for 15 min in hexane, acetone and ethanol, successively, dried at 80 °C for 5 min. The surfaces of the chips were aminosilanized at 80 °C overnight using APTES in dry xylene in the presence of a catalytic amount of diisopropyl- ethylamine as aminosilanizing agent (Lamture et al. 1994). After aminosilanizing the chips were washed thoroughly with ethyl acetate and used immediately.

Immobilizing IgG on the Glass Chips. Immobilization of the IgG was accomplished by first immersing the amino-modified chip in 5% glutaraldehyde in phosphate buffer saline (PBS-T, Na_2HPO_4 8.1mM, NaH_2PO_4 1.9nM, NaCl 0.14M, Tween 20 0.05%, pH 7.4) for about 5 h. Then the chip was rinsed with deionized water and dried in a stream of nitrogen. The 0.2-10 µg/ml rabbit anti-human antibody in carbonate buffer (0.015 M $NaCO_3$, 0.035 M $NaHCO_3$, pH 9.6) was dispensed drop by drop onto the activated surface of the chip by robotic microdispensing system made in our lab. The diameter of each spot was about 120 µm and the distance between two spots was 200 µm. The liquid volume of each distribution is of the order of picoliters. The reaction was allowed to progress at 37 °C for 2 h. Then the glass slide was rinsed with PBS-T three times. The modified substrate was next soaked at 37 °C for an hour in 1% bovine serum albumin (BSA) in 0.1M borate buffer (pH 8.7) to block the residual aldehyde group. Finally, the antibody-coated substrates were rinsed with PBS and dried quickly under a stream of argon.

Preparation of Fluorescent QDs-antibody Conjugate Colloids. Fluorescent QDs-antibody conjugate colloids were prepared in a two-step process. In the first step, the colloidal QDs in chloroform were reacted with 1.0 M glacial MSA aqueous solution for 2 h with vigorous shaking. A PBS solution was added at 1:1 volume ratio. After vigorous shaking and mixing the chloroform and water layers separated spontaneously. The carboxyl-coated QDs, which were distributed in the aqueous layer, were aggregated during centrifugation at 16000 r/min for 12 min. The clear supernatant was discarded. Excess MSA was removed through four or more rounds of centrifugation. Finally, the carboxyl-QDs were homogenously dispersed in the aqueous phase.

The second step of the preparation involved immobilization of IgG on carboxyl-QDs. The carboxyl-QDs were reacted with 1% (w/w) EDC in hydrous acetonitrile for 5 h. The residual EDC and acetonitrile were removed by five cycles of PBS washing and centrifugation at 16000 r/min for 15 min. While gently agitating, antibody (rabbit anti-human) in PBS was added to the 200µL suspension of QD colloids. The mixture was incubated at 4 °C overnight, during which time the antibody immobilized on the colloids. The excess unreacted antibody was removed by several rounds of centrifugation. These suspensions were usually stable for many days when stored at $-20\,°C$.

Immunoassay Protocol. The immunoassays were conducted by following the typical procedure for sandwich-type assays. Human IgG and goat IgG were used as test antigens. In each case, 100 µL of IgG solution in PBS buffer was spread onto the capture antibody-coated slide and allowed to react for 1 h at room temperature. After rinsing with PBS-T three times the slide was exposed to 100 µL of fluorescent QD-antibody conjugate colloids for 3 h. All samples were then rinsed with PBS-T and dried under argon before scanning fluorescence imaging.

The 532 nm laser line was used for excitation of immune spots. The fluorescence intensities of the spots were recorded on the laser confocal scanning microscope and analyzed quantitatively by the Imagene software.

3. Results and Discussion

The Effect of Coupling of MSA on the Fluorescence Properties of the QDs. The luminescent QDs were prepared in trioctylphosphate and trioctylphosphate oxide solvent. It is difficult to disperse them homogeneously in aqueous solution. Furthermore, it is unclear how to attach biomolecules to a QD while still maintaining their fluorescence activities. Chan et al. reported a method of using mercaptoacetic acid for solubilization and covalent protein attachment (Chan and Nie 1998). We used MSA to replace mercaptoacetic acid because one MSA molecule provides two carboxyl groups that may increase the number of IgGs bound to each QD. In addition, the MSA layer is expected to reduce passive protein adsorption on QDs. The fluorescence emission peaks of the carboxyl-QDs and IgG conjugated QDs were at 588 nm and 590 nm, respectively. The results indicated that the optical properties of QDs remained unchanged after coupling and conjugation.

The Advantages of QDs as Biological Probes. The fluorescence spectra of the traditional biological probe Rodamine 6G and QDs are presented as Fig. 1a and Fig. 1b, respectively. Compared with Rodamine 6G, QDs can be exited in a broad wavelength band (250 nm) that has a shorter wavelength than the emission wavelength. Most laser excitations can be used to excite them. Furthermore, QDs emit light in a very narrow spectral range. This allows many different coloured QDs to be used simultaneously without the problem of interference between the different colours. Moreover, the large Stokes shift is the really attractive characteritic of QDs. In contrast, the small Stokes shift of organic fluorophores led to the difficult problem of self-quenching and scattering interference between the excitation and the emission bands.

a)

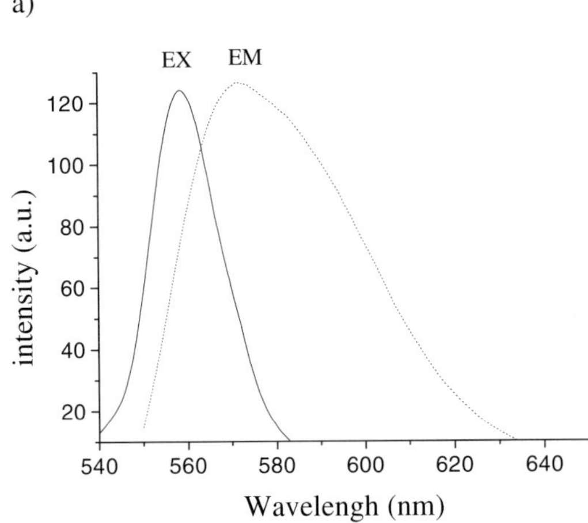

b)

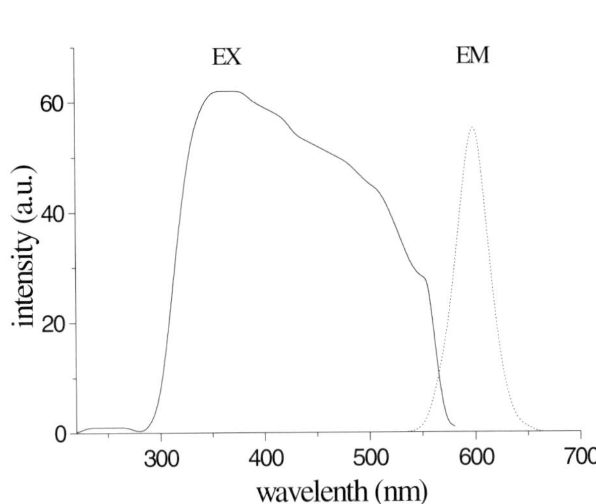

Fig.1. (a) Fluorescence spectrum of Rodamine 6G, **(b)** Fluorescence Spectrum of QD

Fluorescent Imaging of QD-labeled IgG. We used the 532 nm laser line for the excitation. The immunoassay picture obtained on the laser confocal Scanner Array 4000 was presented in Figs. 2 and 3. The rabbit anti-human antibody labeled with QD recognized human IgG specifically, which led to extensive aggregation of the fluorescent QDs. The site where there were immobilized human IgG was bright, while the site without capture-antibody where QD was not absorbed

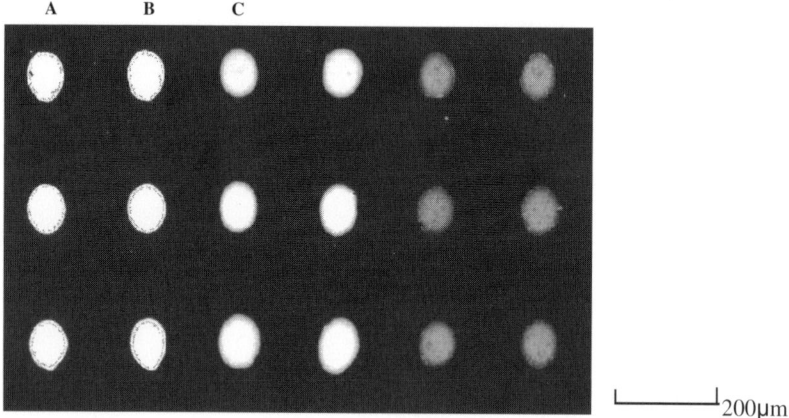

Fig. 2. Detection immunoassay on glass slides. IgG concentrations are succesively diluted from A to C. In all panels, the fluorescence is false-coloured white, green and blue, respectively

Fig. 3a. A part of a single slide with 16×16 spots. All the spots contained the same concentration of IgG

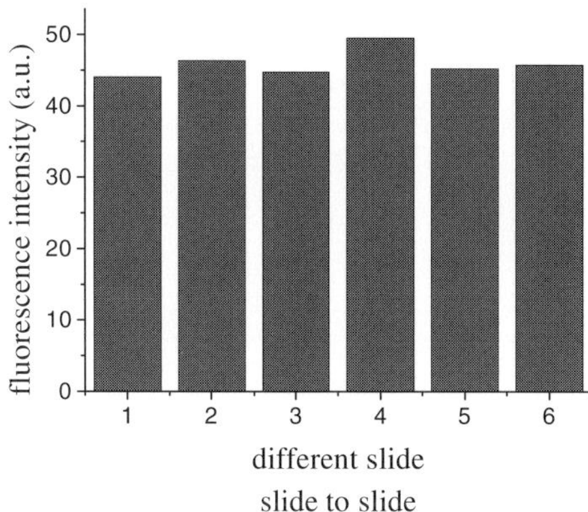

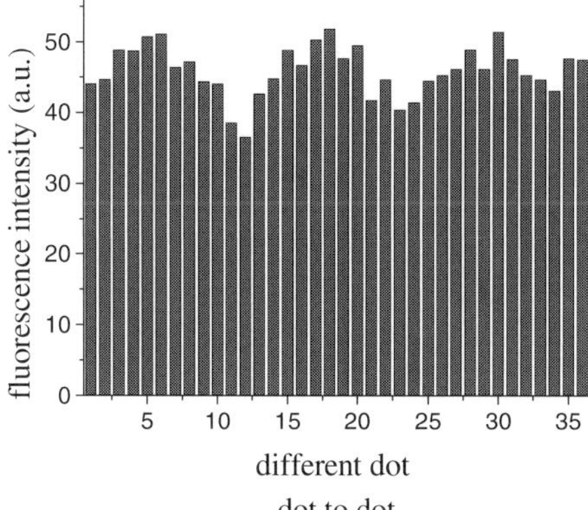

Fig. 3b. The different fluorescence intensities from dot to dot on the same slide and from slide to slide

was dark. In this way the IgG is determined by the brightness given by the scanner, as presented in Fig. 2. When the spots of a glass slide were dispersed with the same concentration everywhere the overall surface of the slide was of almost uniform brightness, as shown in Fig. 3. In contrast, the glass without any immobilized antibody was dark. The experiment showed that the attached immunomolecules could congregate specific antibody or antigen. At the same time, an increase in the antigen concentration resulted in a gradually increasing spot fluorescence intensity, as illustrated by pure human IgG curve in Fig. 4.

Specificity Experiments. Specific recognition of antigenic species in aqueous samples and QD-labeled antibody using sandwich type immunoassay was demonstrated on the slide. These experiments used rabbit anti-human IgG labeled with QDs for immune recognition. Rabbit anti-human IgG immobilized on the glass slide was used as the capture antibody. The mixture of human IgG and goat IgG in different ratios (w/w) was spread onto the capture antibody-coated substrate and allowed to react for 1 h at room temperature. The substrate was exposed to QD-labeled rabbit anti-human IgG solution for 3 h. The fluorescence intensity of the labeled sites increased with increasing weight fraction of human IgG (Fig. 4). This result indicates that human IgG was specifically captured by the substrate and also specifically bound to rabbit anti-human IgG-QD colloids while the goat IgG was not. The result is in agreement with the expected specificity for sandwich assay.

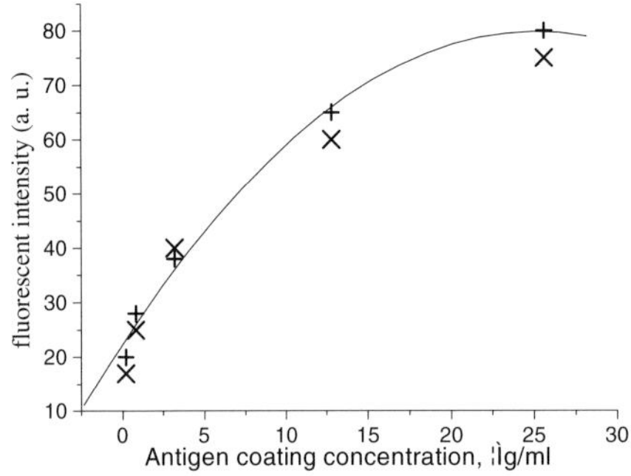

Fig. 4. Dependence of fluorescence intensity on antigen concentration: + = human pure IgG × = human IgG in a mixture of human IgG and goat IgG

4. Conclusion

An alternative immunoassay readout method on microminiaturized glass chips is described based on fluorescence imaging analysis with laser confocal scanning. In principle, the result obtained from the microminiaturized glass chips shows that the readout method developed with QD-labeled biomolecules such as antibodies, antigens or enzymes is well suited to the application on protein chips. Further synthetic efforts may produce sufficiently monodispersed QDs (Peng et al. 1998), which should be an important step toward developing multiplex detection schemes. Experiments along these lines are underway.

Acknowledgement

We express our appreciation of helpful discussions with Prof. Xinrong Zhang. This work was supported by National Science Foundation of China (39989001).

References

Andrew TT, Mirkin CA and Letsinger RL (2000) Scanometric DNA array detection with nanoparticle probes. Science 289: 1757

Carl AKB (2000) Antibodies in diagnostics – from immunoassays to protein chips. Immunology Today 21: 379

Chan WCW and Nie S (1998) Quantum dot bioconjugates for ultrasensitive nonisotropic detection. Science 281: 2016

Ekins R (1990) Multispot, mutianalyte immunoassay. Ann Bio Clin 48: 655

Lamture JB, Beattie LK, Burke BE, Eggers MD, Ehrlich DJ, Fowler R, Hollis MA, Kosicki BB, Reich RK, Smith SR, Varma RS and Hogan ME (1994) Direct detection of nucleic acid hybridization on the surface of a charge coupled device. Nucl Acids Res 22: 2121

Macbeath G and Schreiber S (2000) Printing proteins as microarrays for high- throughput function detection. Science 289: 1760

Marcel BJ, Mario M, Peter G, Shimon W and Alivisatos AP (1998) Semiconductor nanocrystal as fluorescent biological labels. Science 281: 2013

Margaret AH and Philippe G (1996) Synthesis and characterization of strongly luminescence ZnS-capped CdSe nanocrystals. J Phys Chem 100: 468

Murray CBD, Norris J and Bawendi MG (1993) Synthesis and characterization of nearly monodisperse CdE (E=S, Se, Te) semiconductor nanocrystallites. J Am Chem Soc 115: 8706

Peng X, Wickman J and Alivisatos AP (1998) Kinetics of II-VI and III-V colloidal semiconductor nanocrystal growth-focusing of size distribution. J Am Chem Soc 120: 5343

Taylor JR, Fang MM and Nie S (2000) Probing specific sequences on single DNA molecules with bioconjugate fluorescent nanocrystal. Anal Chem 72: 1979

Trends in Microfluidic Devices for Analytical Chemistry

Zhang C. X. and Manz A.

Abstract

This chapter reviews recent advances and future trends in microfluidic devices for analytical chemistry. Microfluidic devices consisting of microchannel networks have offered an exclusive opportunity for performing massive parallel analysis with an integrated chip of a few cm. The microfluidic technique has yet to be a landmark in analytical chemistry, just like the role of the electronic chip in computer development. This review chapter consists of six sections: (1) introduction, (2) flow control, (3) sample handling, (4) separation, (5) microfluidic devices for electrospray ionisation mass spectrometry, (6) conclusions and outlook. Stress is laid on biomedical applications.

1. Introduction

Analytical chemistry (Harvey 2000) is the science of inventing methods to answer two kinds of questions: (1) what molecules are present (referred to as qualitative analysis covering structure elucidation) and (2) how many molecules are present (quantitative analysis including concentration determination). Analytical chemistry has been playing a vital role in the development of many sciences. In history there have been numerous examples in which revolutionary progress in an individual science subject was initiated by an invention of analytical technique. The well developed modern analytical techniques, including optical spectroscopy (absorption, emission and fluorescence spectrometry), nuclear magnetic resonance spectrometry, mass spectrometry and chromatography (gas chromatography, high performance liquid chromatography and electrophoresis), have led to entirely new insights into life processes on the molecular level rather than on the species behavioural or cellular level . A new type of knowledge for better understanding of life phenomena from the molecular point of view can be obtained with bioassays (Gelashvili et al.1999), i.e., determination of endogenous and exogenous substances in biological matrices. While human beings have conquered the air, we know very little about ourselves. There have been increasing demands for molecular information about our bodies: DNA, RNA, proteins, etc. New pharmaceutical products are increasingly required to combat diseases. Consequently, improved analytical methods in terms of high throughput, small sample size, small set-up size, low costs and easy disposal of the wastes are required.

To take an example, there is a new tendency in drug discovery: the use of combinatorial chemistry (Terrett 1998). In drug discovery, all related structural variants of a given candidate are simultaneously synthesized producing a large number of compounds, i.e., a combinatorial compound library, so that no compound is missed. In order to find the right substance, each compound in the combinatorial library is tested against a target of diseases with high throughput screening techniques. Combinatorial chemistry has led to a greatly accelerated drug discovery process compared to the traditional one-compound-at-a-time approach. It is because of the advances in high throughput analytical techniques that combinatorial chemistry has become practical. The combinatorial approach (i.e., multiple processing) has also been used in high throughput screening](Todd et al. 1999) and optimization of various chemical (Sullivan et al. 1999) or biochemical processes (Ma et al. 2000).

To keep pace with the high throughput combinatorial synthesis, parallel analysis, i. e., simultaneous multi-sample analysis, is a natural solution. However, not every technique can easily be paralleled. For example, despite high performance liquid chromatography (HPLC) still remaining the most popular separation analytical technique for bioassays performed in sequential mode, it is very difficult to build a highly multiplexed HPLC system because of its bulky column size (a few mm in outer diameter). Two HPLC columns are probably close to the practical limit that can be incorporated into one set-up, with which the number of compounds that can be analyzed in a given time period can be doubled at most (Zeng and Kassel 1998). For similar reasons, the number of packed gas chromatography columns that can be multiplexed is limited to a few (Prazen et al. 1999). Obviously they cannot cope with the high throughput of combinatorial compounds generated rapidly with the use of 96-well plates.

The much smaller size of capillaries (150-350 µm in outer diameter) employed in capillary electrophoresis (CE) makes it relatively easy to parallel. In a multiplexed CE scheme reported recently, a total of 96 capillaries (150 µm in outer diameter) are packed side by side at detection windows producing a combined width of 15 mm (Gao et al. 1999; Gong and Yeung 1999). The optical signals are registered with a 1024-diode photodiode array (Gong and Yeung 1999) or a 5x300-pixel CCD camera (Gao et al. 1999). At the injection end the 96 capillaries are fitted into a 96-well microliter plate that is often used in combinatorial synthesis. The 96-capillary array seems ideal for combinatorial screening and its high throughput has been demonstrated in comprehensive peptide mapping (Kang et al. 2000), enzyme activity evaluation (Ma et al. 2000) and DNA mutation screening (Gao and Yeung 2000). Driven by the Human Genome Project (Venter et al. 1998), several groups have been devoted to developing capillary array electrophoresis (CAE) for DNA sequencing (Anazawa et al. 1996; Ceriotti et al. 2000; Huang et al. 1992) which will finally lead to the routine use of a commercial high throughput sequencer based on a 96-capillary array (Venter et al. 1998). Therefore DNA sequencing can be performed more quickly and at lower cost than previously anticipated and it has been predicted that the genome would be completed well ahead of 2005, the target completion date (Waterston and Sulston 1998). An even more powerful multiplexed system incorporating 384 capillaries has been commercialized. However, as the number of capillaries in CAE increases, manufacturing and manipulation become increasingly difficult, giving rise to further advanced techniques of higher throughput being required.

Recently, microfluidic devices provided a potentially highly promising alternative to CAE for high throughput analysis. A group of 96 separation channels (equivalent to 96 capillaries) has been fabricated in a 10 cm-diameter glass substrate (Shi et al. 1999). The distance between channels can be as small as 90 µm (Woolley et al. 1997). Because all the channels are micromachined into a single piece of substrate, it is much easier to handle than side by a side bonded capillary array. Since CE was demonstrated in microfabricated channels of 30×10 µm in 1992 for the first time (He and Regnier 1998), a burst of interest in microfluidic devices has been triggered. As of today, there have been attempts to transfer almost all the major bench-top analytical approaches to the microfluidic chips, including separation (Kutter 2000), optical detection (Salimi-Moosavi et al. 2000) and the electrospray ionization interface for mass spectrometry (Oleschuk and Harrison 2000). The diversity of molecules addressed by microfluidic devices is large, from small amino acids (Effenhauser et al. 1993), the peptides (Cohen et al. 1999) to macro proteins (Jiang et al. 2000) and DNA (Liu et al. 1999). In evidence, the bi-annual conference Micro Total Analysis Systems dedicated to miniaturized chemical systems has been updated to a once-a-year basis to keep pace with the fast developments (Berg 2000). Also, a microfluidic product for DNA, RNA and protein analyses has recently been marketed by Agilent (Mueller et al. 2000). Dozens of companies are actively involved in research on microfabrication techniques looking for commercial opportunities (Jakeway et al. 2000).

The reason miniaturized systems are gaining popularity is self evident: size matters. A miniaturized system can contain a high density of instrumental components and thus facilitate multiplexed systems for providing high sample throughput. Single analysis can also be drastically accelerated with the decreased setup sizes because distances for molecules to travel are drastically shortened (this will be discussed in detail in Sect. 3). Co-ordinated with the reduced size of the microfluidic devices, the sample size is minimized. For example, in microchip CE in general less than one nL of sample is required, compared to the average 10 µL of sample used in HPLC. This is important because the biological samples often exist in tiny amounts and therefore cannot be analyzed with the 'macro-methods'. As the setup size, sample size and therefore the reagent size are reduced, the costs of analysis using microfabricated devices are reduced as well. More importantly, all the steps that are involved in a whole analytical process, from handling of the real samples to acquisition of signals of the analytes of interest, can be integrated into a single microchip to form a micro total analysis system (µ-TAS). A typical bioassay process, for example measuring a compound concentration in the blood, is a multi-stage process consisting of sample preparation, separation of interfering sample matrix from the compound of interest, identification and detection. Most of the assays reported involve using several separate devices to implement the whole process. To increase the overall throughput and the levels of automation of a method, all of the steps involved must be coordinated and the number of manual steps should be minimized. For this purpose, an integrated microfluidic device is a natural solution. All the functional units involved in an analytical process are incorporated onto a single microchip of a size equivalent to a computer chip. The molecules can be transported from one unit to another through the microchannel network, in analogy to the movements of electrons in the microcircuits of computer chips. Thus, the concept of micro total analysis systems (µTAS) (Manz et al. 1990) was proposed ten years ago and now it has become the major driving force in the development of microfabricated devices.

Microfabricated devices for chemical analysis are generally made with glass, silicon or plastic substrates 0.2-1 mm in thickness and 1-10 cm in diameter using micro-machining techniques similar to those used in the manufacture of micro-electronic devices (Eijkel et al. 1999; Jakeway et al. 2000). The miniaturized chemical analysis systems are generally divided into two groups: microfluidic and micro-well array devices. The latter is simply equivalent to arrays of tubes at the microliter level fabricated in a chip with high vessel-density. A 3456-well assay plate project has been reported, each well holding 0.2-2.2 uL of liquid (Mere et al. 1999). Thousands of chemical or biological processes can be performed simultaneously. Theoretically all types of assays based on signals arising from chemical reactions can be implemented in the microarray devices. The microarray approaches have been successful in DNA hybridization analysis (Cheng et al. 1998; Swanson et al. 2000; Vo-Dinh et al. 1999). Numerous DNA probes of different sequences are immobilized on the surface of a wafer in an array format (so-called DNA biochip). Target DNA samples are exposed to the biochip and selectively hybridize with the complementary probes. The hybridized probes are spotted by scanning the chip surface. Microfluidic devices consist of microchannels (1-100 µm in depth, width or diameter) through which molecules are transported. Microarray and microfluidic approaches are competitive in some aspects and complementary in others. Microfluidic devices are anticipated to be the format of choice for integrated and automated analysis because to complete a total analysis process in a single device transport of molecules through microfluidic channels must be involved.

Compared to the micro-well array technique, which already revolutionized genomics (Marshell and Hodgson 1998), the microfluidic system is relatively new and in its infancy. Here we review general trends and recent developments in analytical techniques based on microfluidic devices. The aim is to draw an anticipated future picture of microfluidic systems rather than to provide an encyclopaedic listing of publications. To address specific subjects there have been comprehensive review papers published recently covering genomic (Ehrlich and Matsudaira 1999) and proteomic analysis (Sanders and Manz 2000), separation methods (Kutter 2000) and the ESI-MS interface (Oleschuk and Harrison 2000). The theory of scaling laws and benefits from miniaturization have been thoroughly discussed elsewhere (Eijkel et al. 1999; Manz et al. 1993).

2. Flow Control

One of the key aspects in implementation of microfluidic processes is accurate and precise control of flows in the network of micro-channels. All processes performed in microfluidic devices, namely sampling, mixing, derivatization, separation and transport of analytes to the detection point, are achieved by means of control of fluids. Using conventional mechanically moving pumps and valves to control pressure-driven liquid flow is straightforward. At an early stage external pumps were employed for delivering samples and reagents to miniaturized systems (Verpoorte et al. 1994). However, the bulky sizes of the pumps defy the conception of miniaturization. Now pumps and valves comprising elastic diaphragms have been fabricated on chips (Eijkel et al. 1999; Jakeway et al. 2000). The motion of the diaphragms is mostly actuated by gas pressures through actuation channels (Jo et al. 2000; Unger et al. 2000). These kinds of pump and valve can be adequate to regulate flow rates at the mL/min level (Jo et al. 2000). However, to regulate liquid flow at nL/min levels, which are commonly encountered in microfluidic methods, a small channel diameter is used and thus high back pressures are incurred Unger et al. 2000). Alternatively, the diaphragm can be actuated with an attached piezo disk (Eijkel et al. 1999; Jakeway et al. 2000). Nevertheless, diaphragm-based approaches suffer from complicated fabrication and the moving parts wear out sooner than other stationary components. Thus a better approach is to manipulate flows without involvement of any moving parts.

The pressure drop (P) in laminar flow channels is governed by the Hagen-Poisseuille equation (Ahn et al. 2000)

$$P = \frac{12\ L\ \eta\ Q}{w\ h^3} \tag{1}$$

where Q is the volume flow rate, η the viscosity, and L, w and h are the channel length, width and height, respectively. Thus the control of flow in the microfluidic network can be realized by designing appropriate channel widths but with the same height in different sections of a microfluidic device, which can be easily fabricated. Such an example is shown in Fig. 1A. At the split-off point a passive valve (i.e., a length of narrowed channel) is placed in one of the channels so that the other channels are filled first. The additional pressure drop (P_2) at the point of restriction is given by

$$P_2 = 2\ \gamma\ \cos\theta \left(\frac{1}{w_1} - \frac{1}{w_2} \right) \tag{2}$$

where γ is the surface tension of the liquid, θ is the contact angle and w_1 and w_2 are the widths of the two sections. All types of flow formats can be achieved through channel design based on Eqs. 1 and 2. With the layout shown in Fig. 1A under constant total flow rate different sections are filled sequentially in the order from 1 to 13 and the back pressure at the input is built up step-wise (Fig. 1B). Once the channels have been filled with liquid A, it can be replaced with liquid B sequentially in channel sections from 1 to 13. In this kind of approach flow is actuated precisely and reproducibly without using mechanically moving valves. The drawback is that it lacks flexibility as the flow is determined by the structure, not by operational parameters.

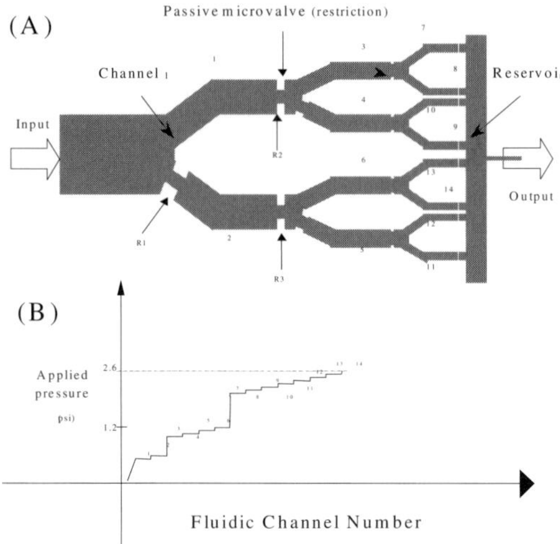

Fig. 1. (A) The layout of a microchannel structure for programmable delivery of liquids. **(B)** Step-wise increases in pressure at the input while the channels are sequentially filled with a liquid at constant flow rate 1μL/min. (Reprinted from (Ahn et al. 2000)with permission.)

Voltage is more popular than pressure in inducing liquid flows in microchip channels because voltage-driven flow is much easier to realize than pressure-driven flow. When a voltage is applied at ends of a channel filled with electrically conductive liquid two types of electrokinetic movements are induced simultaneously: electrophoresis of a specific solute (linear velocity V_e) (Mosher et al. 1992) and electroosmosis of the bulk liquid (V_{eof}) (Thormann et al. 1998) according to

$$V_e \ = \ \mu_e \ E \ = \ \frac{q}{6 \pi \eta r} \ E \tag{3}$$

$$V_{eof} \ = \ \mu_{eof} \ E \ = \ \frac{- \sigma \delta}{\eta} \ E \tag{4}$$

where E is the electric field strength, q and r are the charge and radius of the solute, μ_e and μ_{eof} are the electrophoretic and electroosmotic mobilities, respectively, σ is the surface charge density of the channel walls and δ is the Debye double layer thickness which is determined by the properties of the liquid (pH, ionic strength, dielectric constant and temperature). The net migration velocity of a solute (V_s) is the sum of electrophoresis and electroosmosis (EOF):

$$V_s \ = V_e \ + \ V_{eof} . \tag{5}$$

Generally theelectroosmotic rate is larger than the electrophoretic rate and consequently all cations, anions and neutral analytes move in the direction of EOF. The EOF is a universal phenomena as charges on a solid surface can always be induced when it is in contact with a conductive liquid. The EOF can be manipulated by modifying the properties of the inner channel walls (σ) or by changing the composition of the liquid (δ and η, Eq. 4). There has long been a stereotype practice that the EOF is suppressed by coating the inner walls of fused silica capillaries (Liu and Kuhr 1999).However, the coating does not always work (Bruin et al. 1990). In fact, the EOF can be better regulated by adjusting the buffer pH (Thormann et al. 1998), by addition of some compounds to the liquid (Zhang et al. 2000) or even by pre-flushing the channels with some liquids (Preisler and Yeung 1996). Nevertheless, electrokinetic movements provide a versatile and convenient means for flow control. One may argue that one drawback of using voltages is that the electrokinetic flow is dependent on the chemical properties of the liquids and channel walls. In fact, this would rather be an advantage as it provides additional means for flow control.

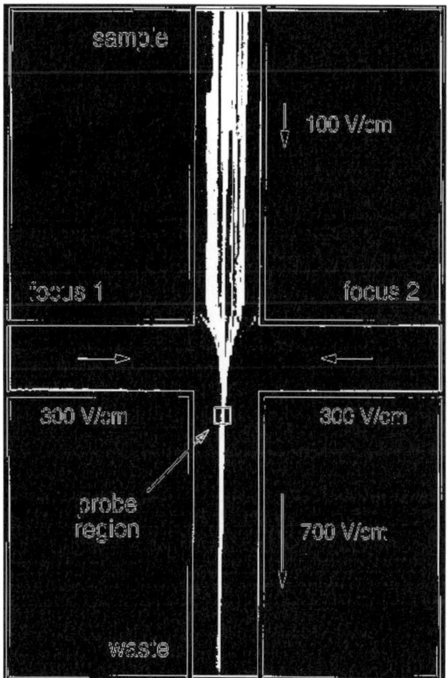

Fig. 2. CCD image of a sample stream (8 µm in width) formed in microchannels 50 µm in width. The focus channels were filled with 10 mM borate buffer (pH 9.2). The sample contained additional fluorescent labelled fluoresbrite microspheres 1.88 µm in diameter. (Reprinted from (Schrum et al. 1999) with permission.)

Figure 2 illustrates the use of voltages to control the flows in a microchannel network. A fine sample flow can be confined at the cross intersection by applying appropriate voltages to each channel (Schrum et al. 1999). The sample stream width is proportional to the ratio of the electric field strength of the sample channel to the waste channel. Confinement of sample flow is required in cytometry and when avoiding contact with channel walls is necessary. Although the sheath flow which encases the sample stream can also be generated with pressure (Tashiro et al. 2000), it seems that voltage provides a better focusing effect. When necessary, the sample flow can be completely blocked by increasing the focusing voltages or decreasing the sample voltage. The so called electrokinetic pumps and valves play a vital role in microchip capillary electrophoresis (see Chap. 4).

Another type of approach for producing valve functions employs polymers that can undergo reversible changes in volumes in response to changes in temperature, liquid pH, etc. (Linden et al. 2000; Liu et al. 2000). Under certain physical or chemical conditions the polymers immobilized in the channel are swollen and thus block the flow while under other conditions the collapsed polymers allow passage of liquid. In a related approach a small plug of liquid is frozen and thawed to close and open the valve. These schemes have wisely utilized the properties of the liquid itself but suffer from long response time (tens of seconds). Combination of pressure and voltage seems an ideal means in liquid control. Caliper is developing so-called universal microfluidic chips in which the flows are manipulated by programming both pressures and voltages at each sample and reagent well (Kopf-Sill 2000).

3. Sample Handling

In typical bioassays, prior to real measurement often dilution or concentration of real-world samples, removing sample matrix, reacting with reagents or amplifying the numbers of copies of molecules are required. For these purposes efficient and fast mixing of two or more liquid flows that join into a single channel is crucial. As laminar flow conditions are always fulfilled in microchannels (the Reynolds number is far less than the threshold 2000, normally less than 10) diffusion is the only mechanism of mixing. Efficient and fast mixing can be achieved by physically splitting the fluid streams into a number of smaller streams as the diffusion distance decreases. The diffusion time, i.e., the time when homogeneity is reached, is inversely proportional to the square of the number of resulting laminae (Ericson et al. 2000). A micromixer has been developed based on this theory, (Fig. 3A) (Bessoth et al. 1999). The infusing flow A is divided into 16 smaller streams of 20 μm each in width in the repeatedly split channels. Flow B is introduced in the same way and joins the 16 laminae of flow A at a confluence point before the sandwiched laminae combine repeatedly into one. Note that the channels for flow A and B are fabricated on different layers of the chip and once they join they flow on the same layer. When the 32 laminae of flows A and B travel to the long channel each lamina is compressed to 5 μm thickness and sandwiched by two laminae of the other liquid except the two outmost laminae. Here fast mixing is initiated.

Figure 3B shows that flourescein and buffer are mixed to 95% of completion in 15 ms. At the channel joining area laminae of flourescein and buffer flows are neighbours of each other. Diffusion occurs rapidly after all the 32 laminae combine into the long channel.

The micromixer has been successfully used for determination of chromium with chemiluminescence detection (Xu etal.2000). To perform chemiluminescence detection, fast mixing of reagents and samples and immediate (almost simultaneously) detection of the signals arising from the chemical reaction are required. Such requirements are perfectly fulfilled with the micromixer shown in Fig. 3. Another application is continuous flow synthesis (Mitchell et al. 2000). The reagents are infused into the system from the two inputs and the reaction time is determined by the length of the long channel. The products can be on-line transported from the output into a mass spectrometer for characterisation. Multiple component synthesis and on-line characterisation can be achieved by continuously infusing liquids composed of various reagents. The micromixer is of great potential power in continuously producing and immediately characterising combinatorial compound libraries.

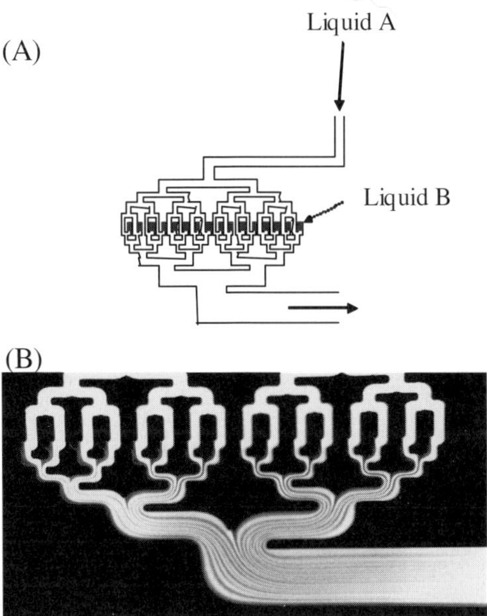

Fig. 3. (A) Schematic diagram of a micromixer (layer 1 for infusing liquid A). The structure for infusing liquid B (same as for liquid A) is on the other side of the chip (not shown). Liquid B meets Liquid A in layer 1 via a number of through-wafer nozzles. **(B)** Flow visualisation of flourescein (in white) and buffer (in dark) at a flow rate of 25 µL/min each in the micromixer obtained with a CCD camera

Careful examination of the performance of the micromixer shown in Fig. 3 reveals that a high percentage of the mixing takes a relatively long time (for example, 98% completion takes 60 ms) (Bessoth et al. 1999). There are three reasons for this: boundary effects, the parabolic profile of pressure-driven flow and corner effects. Boundary effects means that the laminae adjacent to the channel walls have only one neighbouring lamina of the other liquid and thus take a longer time to reach homogeneity than the rest of the laminae that have two layers of the other liquid on both sides. Due to the characteristic parabolic profile of pressure-driven flow (Holm et al. 2000), fluid layers in the centre of the channels move faster than those close the channel walls. The corner effect stretches the front of a flow when it traverses a turn of a channel. The molecules adjacent to the inner walls migrate a shorter distance than the molecules close to the outer walls and thus pass the turn first, leading to a skewed flow profile. This phenomenon can be clearly seen from the simulation data shown in the upper panels of Fig. 4 (Molho et al. 2000). The data shown in Fig. 4 are for electrokinetic flow, but hydrodynamic flow suffers from the same thing. The turn-induced problem can be solved by reducing the channel width at the turn (Fig. 4, lower panels) (Molho et al. 2000; Paegel et al. 2000).

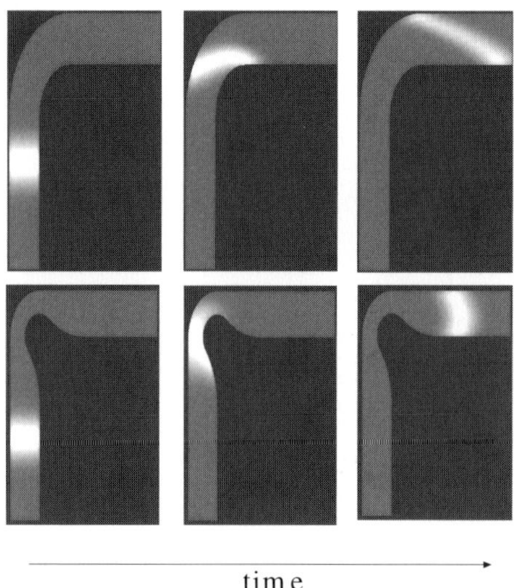

Fig. 4. Simulation of a liquid band travelling through a constant radius corner (upper panels) and a tapered turn (lower panels) driven by voltages (Reprint from (Molho et al. 2000)with permission).

Mixing two fluids in different ratios can be easily realized using voltages and a simple T-intersection where two input channels and one output channel join, if the mixing time is not important (Harrison et al. 1993). An increased voltage at the reagent reservoir leads to a further diluted sample solution in the outlet channel. Based on similar principles, a serial mixing device was developed recently where the sample was sequentially diluted in the subsequent cross intersections (Fig. 5A)

(Jacobson et al. 1999). Sample from channel S_1 is split into the channels A1 and S_2 at the first cross intersection and the aliquot entering S_2 is combined with buffer flow from channel B_1. Part of the diluted sample in channel S_2 is directed to A_2 for analysis and the rest is further diluted in the second cross intersection. Consequently, a series of diluted sample solutions (or ratios of sample and reagent) are produced in channels A_1 to A_5. The dilution factors are determined by the voltages applied at the buffer and sample reservoirs and the layout of the mixer. A distinct boundary between sample and buffer can be seen just after the cross intersection (Fig. 5B), while homogeneity is achieved when the mixture arrives just before the next cross intersection. However, when the voltages at the buffer and sample reservoirs increased from 0.4 to 1.6 kV the liquids travel too fast to reach homogeneity before the next mixing step takes place (Jacobson et al. 1999).

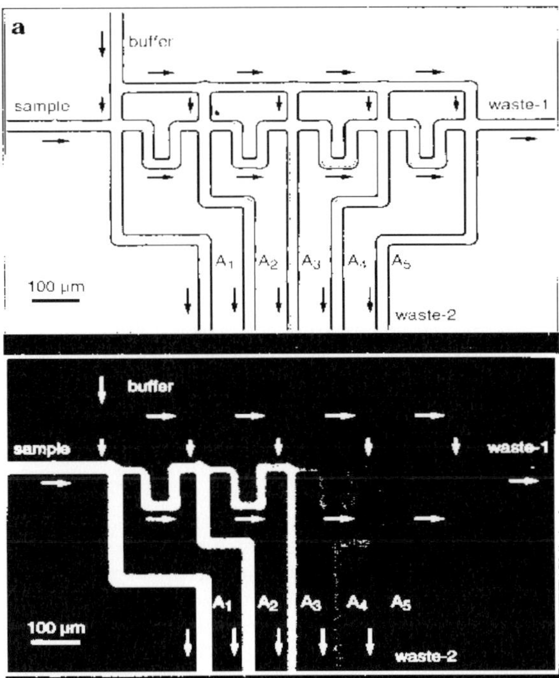

Fig. 5. (A) White light image of a serial mixing manifold. **(B)** CCD image of serial mixing of rhodamine sample solution (in white) with buffer (in dark) using a voltage of 400 V at the buffer and sample reservoirs while the waste wells are grounded. The channel width is 24 µm and the length between adjacent cross intersections is 0.4 mm. Reprinted from (Jacobson et al. 1999) with permission.

Another type of micromixer is based on use of a piezo disk that is mounted on the silicon membrane of a micro-chamber (Woias et al. 2000). Mixing is achieved by shock waves which are in turn achieved by the vibrating membrane. This system is suitable for large flow rates (95-490 µL/min).

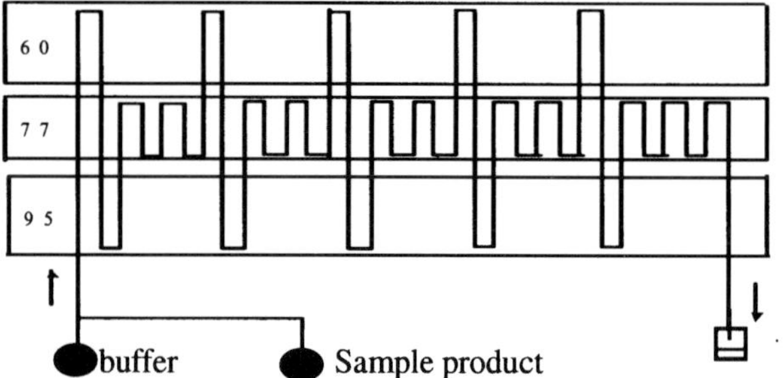

Fig. 6. Schematic diagram of a continuous flow PCR microfluidic device (5 identical cycles are shown). Three copper blocks, which are heated by 5 W heating cartridges and mounted on the glass chip, create three distinct temperature zones. Channel dimensions: 40 µm x 90 µm. The total channel length for 20 cycles is 2.2 m. In each cycle the channel length ratio for the 95°, 65° and 77° C zones is 4 : 4 : 9, equivalent to the time ratio for melting, annealing and extension. The arrows indicate the flow direction.

Often in biological and environmental samples the analytes of interests are of too low concentration to detect. In these cases chemical amplification can be carried out in a way analogous to an electronic amplifier. For example, in DNA analysis the number of specific DNA molecules is doubled in a polymerase chain reaction (PCR) cycle (Mullis et al. 1994). After n PCR cycles the total number of DNA molecules is amplified by a factor of 2^n. Each PCR cycle involves three steps (melting, annealing and extention) that are realized at three different temperatures (95, 60 and 77°C, respectively). PCR is conventionally performed by repeatedly heating and cooling a reaction chamber. The efficiency and time, which are largely determined by the thermal conductivities of the container and the reaction liquid, can be improved by miniaturizing the reaction scale. According to a recent report, cycle times can be shortened from 3 minutes to 30 seconds when the PCR is performed in microchip chambers of 0.28 µL instead of conventional containers of 10-25 µL (Lagally et al. 2000). A continuous flow PCR chip based on a new scheme of temperature control has been developed (Kopp et al. 1998). The PCR liquids flow continuously in a microchannel that passes repeatedly through three temperature zones (Fig. 6) and therefore repeatedly experience temperature cycles. The reaction time at each temperature is determined by the channel length in the temperature zone and the flow rate. Using a continuous flow PCR chip of 20 cycles the PCR mixture can be hydrodynamically driven through the whole channel experiencing cycle times of 4.5 to 50 s if the flow rate ranges from 4.3 to 0.3 µL/min. Small amounts of DNA sample can be injected into the chip and amplified.

Not all analytes can be chemically amplified. A more common type of approache addressing low concentration samples is to perform sample preconcentration. Traditionally, sample preconcentration and the removal of sample matrix are accomplished using off-line liquid-liquid and solid phase extraction, which are time consuming, labour intensive and often the bottleneck of the whole analysis process against high sample throughput. Incorporating solid-phase extraction in microchips has just been reported for the first time (Oleschuk et al. 2000). The preliminary results show that BODIPY can be concentrated by a factor of 100 using a micro-chamber of 300 pL packed with porous C-18 beads 1.5-4 μm in diameter. It is interesting to note that voltages were used as the driving force even in the elution step with pure acetonitrile. In another approach to sample concentration, a thin sodium silicate bonding layer between the two glass plates is used as a membrane connecting two channels distanced by 3-12 μm (Khandurina et al. 1999).When a voltage is applied between the two channels, the membrane allows passage of small ions (salts) to establish electrical connection while preventing large DNA segments from traversing the physical barrier. Up to two orders of magnitude concentration factors have been obtained in this approach. Due to the small cross-sectional area of the membrane compared to that of the channel, however, most of the voltage potential drop is over the membrane and leads to a low electric field strength along the channel. Consequently, the DNA ions were found to move slowly and a long preconcentration time (250 s) was required.

An on-line sample preconcentration method based on Ohm's law field-amplified sample stacking (FASS) has been well explored in capillary electrophoresis (CE) (Zhang and Thormann 1998) Samples can be concentrated three order of magnitude by simply preparing sample solutions of low conductivity. Although FASS has not been widely appreciated in the designs of microfluidic devices, its power has been demonstrated in a few microchip CE approaches where the concentration sensitivity can be easily improved 3-50 fold (Lichtenberg et al. 2000; Li et al. 2000) .Another very interesting electrokinetic preconcentration approach (Liu et al. 2000) shows that rhodamine B can be concentrated by a factor of 560 using micelle sweeping (Quirino et al. 2000) and transport-number mismatch mechanisms. Nevertheless, electrokinetic sample preconcentration has yet to play a key role in microfluidic system.

A microfabricated dual-microdialysis device has been developed for removing sample matrix (Xiang et al. 1999). Two dialysis membranes (molecular weight cut-off 50 kDa and 8 kDa) are sandwiched between three substrates with four micromachined channel layers. Dialysis membranes allow molecules smaller than the cut-off size to pass through while blocking the large ones. Cell lysates are infused by a syringe pump into the first channel layer. A portion of sample diffuses through the high molecule membrane into the second channel layer, leaving cellular debris and high molecular mass components (>50 kDa) in the first channel layer flowing towards the waste exit. The sample entering the second channel layer is guided to the third channel layer through a drilled hole and exposed to the low molecular weight membrane. Small molecules like inorganic salts (<8 kDa) diffuse into the fourth channel layer through the low molecule membrane. The clean sample (clear of both high and low molecular weight components) left in the third channel layer is directly transported into an electrospray ionisation mass spectrometer (ESI-MS) for characterisation. Crude samples can be on-line pretreated and analysed by ESI-MS (Xiang et al. 2000). Recently, filter structures have been fabricated on microchips. Arrays of micro-posts spaced 1.5-7 μm apart across the microfluidic channel serve as filters (He et al. 1999; Wilding et al. 1998). Alternatively, a microfilter can be made using weirs and the gap between the tops of channel dams and the cover plate serve for filtration. The microfabricated filters at μm levels are specially useful for isolation of cells with μm sizes as well (Wilding et al. 1998). For isolating molecules, dialysis membrane-like structures with pore sizes at sub-nm levels are required, which is a challenge to currently available micro-fabrication techniques.

4. Separation

Separation plays a central role in μTAS as most complex biological samples cannot be determined without separating the analytes from each other and from interfering components. Figure 7 shows the principle of analytical separation techniques including gas chromatography (GC), high performance liquid chromatography (HPLC) and capillary electrophoresis (CE). At the beginning a small sample plug is injected into the separation channel at the inlet and then the separation process starts by pumping liquid or gas flow through the channel carrying the sample components (analytes) towards the outlet to a detector. Due to the different physical and chemical properties of analytes they migrate at different velocities and therefore arrive at the detector at different times. The time for an analyte to reach the detector (the so-called retention, elution or migration time) is specific for a given set of separation conditions and can be used for identification. The peak height or area is proportional to the analyte concentration and is employed for quantitative analysis. The peak width represents separation efficiency. The narrower the peak (analyte band), the higher the separation efficiency and the larger the theoretical plate number. The later parameter is used for quantitative evaluation of separation efficiency.

Plenty of knowledge on HPLC obtained in the last half century and a large number of HPLC experts have naturally prompted the motivation to implement this popular bench-top system in a microfabricated format. Actually, the first HPLC microchip was fabricated 10 years ago (Manz et al. 1990), well ahead of the first CE microchip (Manz et al. 1992). In HPLC separation, the theoretical plate number N is given by

$$N = \frac{L}{2 \lambda D} m \tag{6}$$

when longitudinal diffusion dominates the band broadening (which is normally true if the flow rate is not very high and the particle size is not very large). Here D is the diffusion coefficient of an analyte in the mobile phase, λ is the hindrance factor dependent on the characteristics of packing, m is the overall migration velocity of the analyte in the column and L is the column length (Braithwaite and Smith 1996). Based on the Kozeny-Carman equation (Krejci et al. 1975), the velocity of an analyte in a packed column is proportional to the pressure difference between the column ends (P) according to

$$m = \frac{P \, d_p^2}{180 \, \eta \, L} \cdot \frac{\varepsilon^2}{(1 - \varepsilon)^2} \cdot \frac{1}{1 + k} \tag{7}$$

where k is the capacity factor of the analyte, η is the viscosity of the mobile phase, d_p is the diameter of the particles packed and ε is the porosity of the packed column. Therefore, the analysis time t (when the analyte migrates from the injector to the detector) is

$$t = \frac{L}{m} = \frac{180\,\eta}{P}\,\frac{L^2}{d_p^2}\,\frac{(1-\varepsilon)^2}{\varepsilon^2}\,(1+k). \qquad (8)$$

Combination of Eqs. 6, 7 and 8 yields the plate number generated per unit time:

$$N/t = \frac{\varepsilon^4}{64800\,\eta^2\,\lambda\,D\,(1-\varepsilon)^4\,(1+k)^2}\,\frac{P^2\,d_p^4}{L^2}. \qquad (9)$$

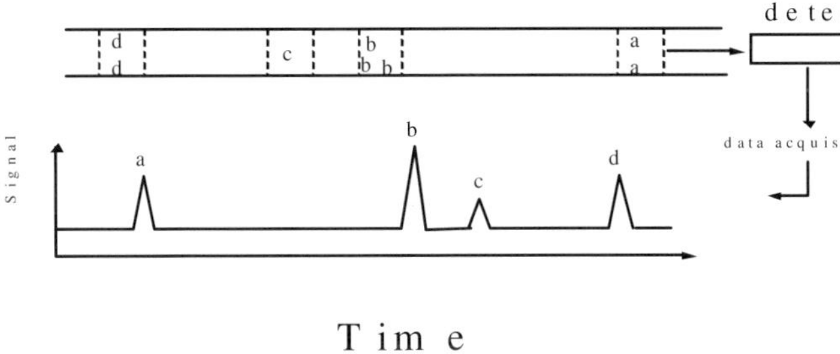

Fig. 7. Principle of chromatographic and electrokinetic separation techniques. Letters a, b, c and d represent four different analytes and the numbers of letters in each analyte band imply the relative concentration of analytes. The arrow in the separation channel indicates the direction of the solute's migration.

To maintain high efficiency, column diameter to particle diameter ratios should stay at certain levels. In miniaturized HPLC system, column and particle sizes should be downscaled simultaneously. It can be seen from Eqs. 6-9 that for a given separation efficiency, if the separation channel (both diameter and length) and particle sizes are downscaled by 10 fold the analysis time is reduced by a factor of 100. This is great as the sample throughput can be increased by drastically speeding up sequentially individual runs, in addition to parallel analysis. However, to achieve such a high speed of separation and efficiency, a 100 times higher pressure is required. In a report using a capillary of 33 µm inner diameter packed with particles of 1 µm, a 5000 bar pressure was required to achieve 20 000 theoretical plates in 6 min (MacNair et al. 1999). The high pressure requirement makes huge technical difficulties in implementation of HPLC in miniaturized systems that must withstand such tremendously high pressures without leaking. On the other hand, if the pressure remains the same, the separation efficiency is worsened by 100 fold and no improvement in analytical speed is obtained compared to the large system. In open tubular HPLC there is the same dilemma: extremely high pressure is definitely needed but is extremely difficult to realize. This is the reason why since the first miniaturized HPLC system was fabricated 10 years ago, no performance data ever followed.

A recently published approach shows that compounds were retained in an open (not packed) microchannel (100x5.3 µm) coated with n-octyltriethoxysilane but no separation was observed (McEnery et al. 2000). As of today, only one paper can be found reporting HPLC separation in miniaturized systems (Ocvirk et al. 1995). The microchannel of 0.1x0.3x20 mm was fabricated with silicon/glass substrates and packed with Nucleosil-100-5-C8 particles (diameter 5 µm). Fluorescein and acridine have been successfully separated. However, the theoretical plate number was only 80, in comparison with 10 000 plates routinely achievable with conventional HPLC systems. Obviously the difficulties associated with packing the channels with beads also contribute to the slow progress in microchip HPLC and other chip approaches involving packed columns (solid phase extraction). To address the packing problems, particle-like supporting of 5x5x10 µm cubes distanced by   1.5 µm have been micromachined in a quartz wafer, resulting in a mimic packed column consisting of numerous microchannels of 1.5x10 µm (He and Regnier 1998; He et al.1998). Although the micromachined support architecture boasts many advantages over packed columns (high degrees of symmetry and regularity), no HPLC separation data has ever been reported. Interestingly, this microsystem originally made for HPLC has been demonstrated in capillary electrochromatography (CEC) separation of peptides (He et al. 1999). To mimic the real functions of packed columns, the distance between the support cubes should be at the nm level, which is a challenge to micro-fabrication techniques. Nevertheless, the difficulties arising from the high pressure requirement, integrating the bulky and cumbersome pumps and valves into a µTAS are currently defying the ambitions of HPLC-lovers. In contrast, open tubular gas chromatography (GC) obviates the difficulties associated with packing and high pressure. Actually, the first report on microchip GC was published as early as 1979 [129]. Various GC components have been fabricated on chips (Lehmann et al. 2000) and a prototype of integrated GC systems based on microfluidic chips has been on exhibition at the µTAS 2000 conference (Frye-Mason et al. 2000). However, the application scope of GC is limited to volatile substances precluding analysis of macro biomolecules.

Compared to the pressure-driven separation techniques (HPLC and GC), voltage-driven counterpart capillary electrophoresis (CE) is the method of choice in miniaturized systems for liquid phase separation. A general trend in separations on microfluidic devices has been the dominance of CE over HPLC. The main reason is that CE is much easier to implement as no pressures, pumps and valves are used (no mechanically moving parts). The flat velocity profile of the voltage driven liquid flow is superior to the parabolic profile of the pressure-driven flow in HPLC, leading to higher separation efficiency in CE. Therefore, the first commercial microfluidic instrument is based on CE separation of DNA, RNA and proteins (Mueller et al. 2000).

The separation in CE takes place in a fused-silica capillary of inner diameter 10-100 μm with an electric field applied between the ends of the capillary (Heiger 1992). The analyte band broadening is dominated by longitudinal diffusion under ideal conditions (Huang et al. 1989). Eq. 6 is valid for CE by taking $\lambda=1$ for open column and substituting V_s (Eq. 5) for m:

$$N = \frac{(\mu_{eof} + \mu_e)}{2\,D}\,E\,L. \tag{10}$$

The analysis time is given by

$$t = \frac{L}{V_s} = \frac{L}{(\mu_{eof} + \mu_e)\,E} \,. \tag{11}$$

Combination of Equations 10 and 11 yields

$$N/t = \frac{(\mu_{eof} + \mu_e)^2}{2\,D}\frac{U^2}{L^2} \tag{12}$$

where U is the total voltage applied. From Eq. 12 it can been seen that in CE the separation speed increases 100 fold if the miniaturized system is downscaled 10 fold while the voltage requirement and separation efficiency remain unchanged. In theory efficiency and speed can be improved infinitely by increasing the voltage. In practice the maximum applicable electric field strength (E_{max}) is limited by the rates of Joule heating and heat dissipation in order to avoid over heated separation channels (Zhang and Thormann 1996). E_{max} is given by

$$E_{max}^{\;2} = \frac{C}{\kappa\,d^2} \tag{13}$$

where C is a constant dependent on the heat dissipation efficiency, κ is the electrical conductivity of the liquid in the channel and d is the diameter of the channel. Combining Eqs. 12 and 13 gives

$$(N/t)_{max} = \frac{C\,(\mu_{eof} + \mu_e)^2}{2\,D\,\kappa\,d^2} \,. \tag{14}$$

The maximum analysis speed can be increased 100 fold when the channel size is downscaled 10 fold. Therefore, CE employing capillaries of small diameters, 10-100 μm, is superior to its traditional counterpart gel electrophoresis and has become a versatile tool in biochemical analysis (Krylov and Dovichi 2000). Despite being a relatively new technique compared to HPLC and GC, CE has enjoyed fast development and successful application in the last 15 years. There is no difference in fundamentals between microchip CE and conventional CE itself. The theories developed for capillaries are readily applicable to micromachined channels. Fast developments of microchip CE and conventional CE have taken place almost in the same time period. The channel size of microchannels (10-100 μm in width or depth) is similar to the size of fused-silica capillaries but shorter lengths are normally employed (1-10 cm). The greatest difference between conventional CE and microchip CE is the different sample injection schemes used. In conventional CE, for sample injection the inlet capillary end is placed in a sample vial and the sample can be injected into the capillary hydrodynamically or electrokinetically. Then the inlet end is put in a buffer vial and a voltage is applied for separation. Sample and buffer do not contact outside the separation channel. In microfluidic devices, however, sample injection is achieved by transporting sample solution through the channel network that physically connects the sample and buffer reservoirs all the time.

In the first generation of CE chips a T intersection was designed for sample injection (Harrison et al. 1993). The injection plug length is determined by the injection time at a constant voltage. The recently developed double T (Effenhauser et al. 1993) and cross (Heeren et al. 1996) injection schemes allow injecting sample plugs of geometrically defined lengths. The sample injection length is independent of injection time and rather determined by the spaces defined by the cross intersection or the two T intersections. A schematic diagram of a cross injector design is shown in Fig. 8A. For sample injection, a voltage potential is applied between the sample and waste reservoirs and sample solution is driven filling the sampling channels through the intersection. Then a voltage potential is applied between the buffer inlet and outlet pulling the sample plug at the cross intersection into the separation channels. Due to diffusion and convection flow, leakage of sample solution from the sampling channels into the separation channels during electrokinetic injection may occur if the separation channels are left floating (see Fig. 8B) (Heeren et al. 1996). The same kind of leakage can also occur in the separation phase if the sampling channels are left floating (Ocvirk et al. 2000). To prevent leakage and to obtain a clearly defined sample injection plug all intersecting channels should be under voltage control at all times (Effenhauser et al. 1997). Figure 8C shows that a well defined sample plug is formed during injection by applying appropriate voltages at buffer reservoirs inducing counter electrokinetic flows in the separation channels towards the intersection to balance the diffusion and convection effects. For separation, a high voltage is applied along the separation channel while maintaining a small voltage at the sample and waste reservoirs. This allows pulling a sample plug into the separation channel and at same time pushing sample solution in the sampling channels away from the intersection to avoid spreading of analytes into the separation channel (Fig. 8D). The continuous voltage control injection method so called pinched injection (Liu et al. 1999) using cross or double T injectors has been a common practice to obtain small, well-defined and reproducible injection plugs, which is crucial to achieving comparable separation efficiency with conventional CE but in a shorter time.

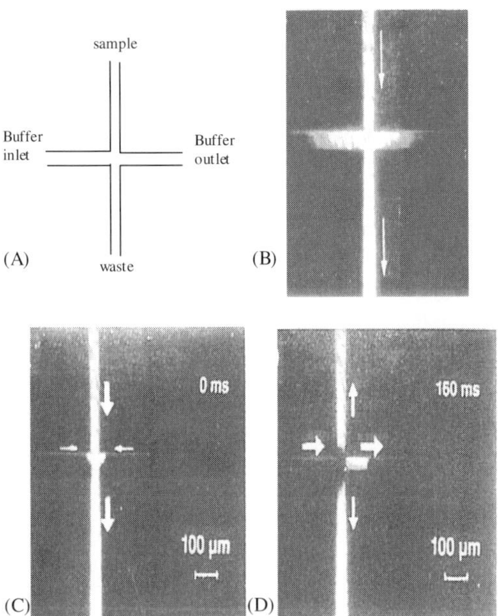

Fig. 8. (A) Schematic diagram of a cross injector. (B, C and D) CCD images during injection and separation. The running buffer was 8% linear polyacrylamide solution containing 100 mM Tris and 20 mM boric acid (pH 9.0). The sample was five fluorescein isothiocyanate labelled amino acids prepared with the buffer. The channel dimensions were 40×10 µm. **(B)** Sample injection was achieved by applying a voltage of 2 kV at the waste reservoir, grounding the sample and **(C)** Floating the buffer inlet and outlet reservoirs. Note that the negatively charged labeled amino acids moved towards the positive electrode due to the week EOF with the polymer buffer (Sample injection with pinch voltages). The voltages at the sample, waste and buffer inlet and buffer outlet reservoirs were 0, 2000, 700 and 700 V, respectively. **(D)** Separation. The voltages at sample, waste, and buffer inlet and buffer outlet reservoirs were 2000, 2000, 0 and 5000 V, respectively (Reprinted from Heeren et al. 1996 with permission.)

Although microchip CE can provide faster separation in comparison with conventional CE, it suffers from lower detection sensitivity (Ramseier et al. 1998). To improve the detection signal larger injection plugs can be obtained using a double T injector instead of a cross one. However, the actual increases in injection size are much less than what are expected based on the relative volumes present in the cross and double-T injectors due to the pinch voltages used. It has been repeatedly reported that only an increase of 20-35% in peak area can be obtained using a 250 µm double-T injector holding a volume five times that of the cross injector (Liu et al. 1999; Wallenborg and Bailey 2000). In practice, the injection plug is determined not only by the injector geometry but also by the voltages used for preventing leakage during injection and separation (Heeren et al. 1996a). Generally speaking, larger injection plugs result in higher signal but lower separation efficiency (Schmalzing et al. 1998). For high separation efficiency an injection plug smaller than the volume defined by the cross intersection can be basically obtained by increasing the buffer flows during injection (Heeren et al. 1996). This allows forming a fine sample stream at the cross intersection (Fig. 2). However, the pinch-injected plug is trapezoidal in shape (see Figs. 2 and 7C), which is a source of band broadening as well. A symmetric and narrower plug can be injected by inserting a step pulling the well shaped plug downstream of the cross intersection back into the intersection prior to the separation phase (Deshpande et al. 2000). A double cross injector has also been proposed for generating narrow injection plugs (Deshpande et al. 2000). Recently, novel multiple sample injection (Kwok and Manz 2000) (several separate sample plugs) and frontal analysis (whole separation channel filled with sample) methods have been developed to improve the detection sensitivity (Kwok and Manz 2002). To register and analyze the complicated signals arising from multiple and continuous sample plugs a special detection scheme known as Shah convolution Fourier transform detection (SCOFT) is used (Crabtree et al. 1999). SCOFT employs multiple detection windows instead of the single-point detection conventionally used in separation techniques. When the analyte traverses the separation channel it is repeatedly registered by the multi-detection windows and therefore the summed signals are enhanced. Basically, SCOFT is applicable in all conventional chromatography and electrophoresis techniques but it is most easily realized in a microfluidic device. After Fourier transform of the original time-domain data, electropherograms similar to that show in Fig. 7 (single sample plug and single point detection) can be obtained but in the frequency domain.

A novel design for gated injection has recently been reported (Fig. 9A) (Jacobson et al. 1999). For loading sample the relay is closed. Because of the electric field strengths in the individual channels determined by the layout, the sample flows through the intersection into waste-1 channel but does not leak into the analysis channel   (Fig. 9B). For sample injection the relay is open for a brief period of time, causing a sample plug to enter the analysis channel (Fig. 9C). Then the relay is closed again to switch off the valve and start separation (Fig. 9D). The injection length is dependent on the injection time (i.e., the time period of opening the relay). With this design only a single high voltage power supply and a high voltage relay are required. Also, the number of reservoirs is reduced. All these features have the potential to simplify the interface between a µTAS and operators.

In CE the maximum plate number generated per time unit is limited (Eq. 14). Further increase in separation efficiency can only be achieved at the cost of analysis time by increasing the length of the separation channel while keeping the electric field strength unchanged (Effenhauser et al. 1993). To increase the channels in a compact device a square cyclic microchip has been developed (Heeren et al. 1996b; Heeren et al. 1996a). With synchronization of a sequence of voltages along different channel sections, analytes can cycle the square loop and be detected repeatedly at a fixed detection point. The resolution between a specific pair of compounds increased as the number of cycles (the equivalent channel length of each cycle is 8 cm) increased (Heeren et al. 1996a). This provides a high degree of flexibility in respect to the choice of the total channel length. However, the plate

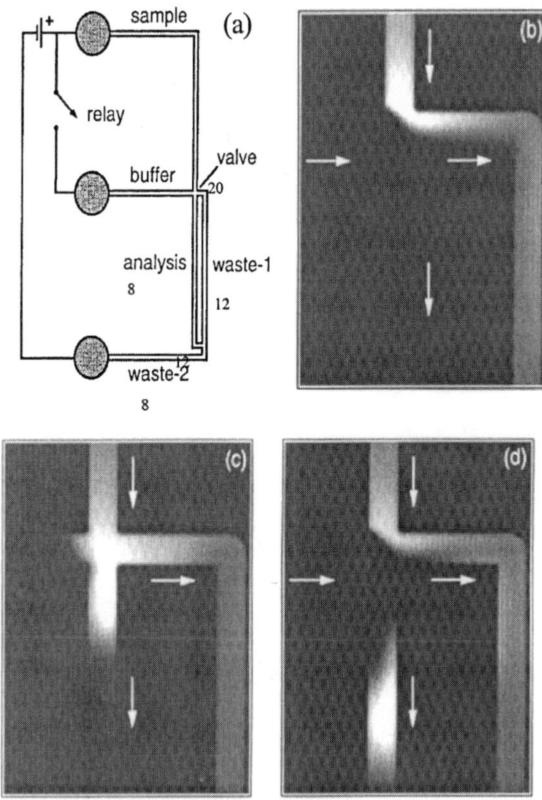

Fig. 9. (A) Schematic diagram of a glass microchip used for gated injection. The channels are of uniform cross sectional area (42x6.9 µm) and the channel lengths in mm are indicated beside the individual sections. (B, C and D) CCD images at the intersection. The buffer was 20 mM borate (pH 9.2) and the sample was 100 µM rhodamine B prepared with the buffer. **(B)** Sample loading. The voltages at the sample, buffer and waste reservoirs were 500, 500 and 0 V, respectively. **(C)** Sample injection. The voltages at the sample, and waste reservoirs were 500, and 0 V, respectively. The buffer reservoir was floating. Injection time: 0.4 s. **(D)** Separation. The voltages were same as for B. (Reprinted from (Jacobson et al. 1999) with permission.)

number does not increase from cycle to cycle in the expected linear fashion. This is due to the corner effect; path lengths around the turns are different. Another kind of approach to increasing the channel length is to use folded channels with U-turns. Using a pair of complimentary U-turns, the turn-induced band broadening at the first U-turn can be corrected more or less by the second one (Culbertson et al. 1998; Paegel et al. 2000). However, a better solution is to use tapered turns (narrowed channel around the turn) (Molho et al. 2000; Paegel et al. 2000). A spiral channel design (i.e., a coiled channel) has recently been proposed to increase the separation length with minimised geometry-induced band dispersion (Culbertson et al. 2000).

In fact, keeping the channel length unchanged, the separation efficiency in CE can be effectively enhanced by control of EOF. In micellar electrokinetic chromatography (MEKC, a mode of CE) the resolution (R_s) is given by (Zhang et al. 1993)

$$R_s = \frac{N^{0.5}}{4} \frac{\alpha - 1}{\alpha} \frac{k}{1 + k} A_{co} \tag{15}$$

where α is the selectivity. Compared to HPLC there is an additional term A_{co}, called the column availability:

$$A_{co} = \frac{L'}{L} = \frac{- \mu_{e,mc}}{\mu_{eof} (1 + k) + \mu_{e,mc} k} \tag{16}$$

Here $\mu_{e,mc}$ is the electrophoretic mobility of the micelles and L' is the virtual column length that a solute really migrates through against the micellar phase. L' is of identical physical meaning as the column length in HPLC. When the micelles are really stationary (i.e., $\mu_{eof} = - \mu_{e,mc}$), Aco = 1, L' = L and Eq. 15 becomes the same as for HPLC. As both the micellar and bulky aqueous phases move, the virtual column length (i.e., the distribution length) can be shorter or longer than the tangible (physical) length depending on the EOF rate. The resolution in MEKC can be improved by increasing the virtual column length while keeping the physical channel length unchanged (Zhang et al. 1993). This is important for microchip analysis as lengthy channels are contrary to the merits of the microfabricated devices. In capillary zone electrophoresis, resolution can be given by (Heiger 1992)

$$R_s = \frac{\Delta\mu}{5.65} \left(\frac{E L}{D (\mu_e + \mu_{eof})} \right)^{0.5} \tag{17}$$

where $\Delta\mu$ is the difference in the electrophoretic mobilities of a pair of solutes. In CZE resolution can also be increased by manipulating the EOF. When $\mu_e = - \mu_{eof}$, R_s is approachies infinity. Although the effect of EOF on resolution is well known, manipulation of EOF for improving separation has not been appreciated in microchip CE. Here manipulation does not mean simply eliminating EOF but rather actively harnessing the EOF.

Simple, short and straight separation channel designs should be employed whenever they can. This is specially required in multiplexed microfabricated devices in respect to separation efficiency and simplified structure (Shi et al. 1999; Simpson et al. 1998; Woolley et al. 1997). The number of reservoirs is two to three times the number of parallel channels, which increases the component density and complexity. Multiplexed microfluidic systems are a particularly attractive approach to addressing the needs of genetic analysis in which large numbers of repetitive measurements are required. A radial microchip consisting of 96 separation channels represents the front line development in paralleled CE analysis (Shi et al. 1999). The radial microplate has a common anode reservoir in the centre of a circular 10-cm-diameter wafer and 96 channels extending outward giving sufficient space for sample and cathode reservoirs. 96 DNA samples can be analysed in 2 min. Various DNA analysis schemes including DNA sequencing (Liu et al. 1999; Salas-Solano et al. 2000) and genotyping (Cheng et al. 1998;Schmalzing et al. 1999; Tian et al. 2000; Woolley et al. 1997] are among the most successful application areas of microchip CE.

The main modes of CE implemented in chip formats remain capillary zone electrophoresis (CZE, for analysis of chargeable substances including amino acids (Hutt et al. 1999), peptides (Cohen et al. 1999), proteins (Jiang et al. 2000) and drug compounds (Ramseier et al. 1998)) and capillary gel electrophoresis (CGE, for DNA analysis). In addition to CZE and CGE, MEKC in chip format has also been proven to be a versatile tool for separation of both charged (Heeren et al. 1996; Wallenborg and Bailey 2000) and neutral compounds (Kutter et al. 1997). Capillary isoelectric focusing (CIEF, another mode of CE) is recognized as a powerful separation tool for ampholytes and especially for proteins (Rodriguez-Diaz et al. 1997). Its chip format has been realized for peptide (Hofmann et al. 1999) and protein (Mao and Pawliszyn 1999; Wen et al. 2000) separation. Capillary isotachophoresis (CITP), a rare mode of CE, has also been implemented in a microfabricated channel of 50x324 μm for separation of potassium and sodium [108].

Capillary electrochromatography (CEC) has emerged as a relatively new voltage-driven separation technique alternative (Smith and Evans 1999). While CEC-fans are happy with the idea that CEC is an independent technique, it is rather one of the CE modes. CEC, CZE, CGE, MEKC, CIEF and CITP can all be implemented using the same experimental setup but employ different fillings (separation media) inside the capillaries. As CEC boasts of superior separation efficiency over HPLC (Ishizuka et al. 2000), CEC has also been pursued in microfluidic systems. Although high pressure is not required, microchip CEC still suffers from the same technical difficulties in channel packing as other packed channel-based microchip techniques (solid phase extraction and HPLC). It has recently been reported that using a tapered channel at the injection end (Ceriotti et al. 2000) or a two-weir design (Ericson et al. 2000) eliminates the need for cumbersome frit fabrication and thus facilitates the packing procedure. However, no decent separation data in packed mcirochannels have been reported. Alternatively, particle-like structures in the separation channel can be formed by micromachining numerous support cubes (He and Regnier 1998; He et al. 1998) or in situ polymerization (Ericson et al. 2000; Fintschenko et al. 2000). With the latter two schemes of column preparation fairly good but not impressive separations of neutral Liu et al. 2000) and charged analytes (Fintschenko et al. 2000;He et al. 1999) were obtained. Compared to the packed column CEC, open tubular CEC using coated capillaries can be easily realized in a microchip format. Separation of four coumarin dyes was achieved by CEC in coated microchannels (Kutter et al. 1998). However, the quality of the separation was no better than that obtained with MEKC using the same chip layout but uncoated channels (Kutter et al. 1997). In fact, open-channel CEC is no different from CZE using coated channels in the perception of an operator. Despite its superior separation efficiency over HPLC, CEC could hardly compete with CZE and MEKC in respect to separation power, not to mention the less critical operating conditions in the latter two.

A new tendency in CE chip development is that plastic materials instead of glass wafers are being used in the fabrication of disposable chips to reduce the costs. Poly (dimethyl siloxane) (PDMS) seems an attractive alternative to glass with comparable performance (Duffy et al. 1998; Effenhauser et al. 1997; Ocvirk et al. 2000). DNA and peptides have even been separated with a PDMS chip (Effenhauser et al. 1997). Other plastic materials that have been used for microfluidic devices include polystyrene and poly(ethylene terephthalate glycol) (Barker et al. 2000).

5. Microfluidic Devices for Electrospray Ionisation Mass Spectrometry

Detection of analytes is the real goal of the whole analytical process. So far, fluorescent emission remains the most popular detection method employed in microfluidic systems due to its simplicity and high sensitivity. However, the majority of analytes possess no native fluorescent emission and a fluorescent reagent has to be tagged to the compounds, which introduces an additional step (derivatization). The simpler UV absorbance detection suffers from low sensitivity associated with small sample loading and difficulty in fabrication of UV transparent microfluidic devices. A recent report has led to a hope that UV detection in microchip would become as versatile as it has been in conventional CE and HPLC (Nishimoto et al. 2000). Indirect fluorescence detection, which is achieved by adding fluorescent substances to the background buffer, can be used as a universal detection scheme but of low sensitivity (Sirichai and Mello 2000). Somewhat specific detection techniques, for examples chemiluminescence (Hashimoto et al. 2000) and amperometry (Wang et al. 2000), have been used in microchip CE. Detection cells for electrochemiluminescence (Arora et al. 1997)and potentiometry (Tantra and Manz 2000) have been fabricated on chips and will possibly play important roles in future μTAS. All the detection schemes mentioned above are for liquid phase. For detection of analytes in gas phase an optical emission detector based on direct current helium plasma has been realized in a chip format and looks forward to integration with microchip GC (Eijkel et al. 1999). Other detection schemes for microchip GC include thermal conductive (Lehmann et al. 2000) and surface acoustic wave (Frye-Mason et al. 2000) detectors. Nevertheless, mass spectrometry (MS) represents a universal and characteristic detection method for compounds existing in both gas and liquid phases. Electrospray ionization mass spectrometry (ESI-MS) is known as a powerful tool for analysis of various chargeable compounds including large proteins and DNA fragments in continuous liquid flow mode (Cole 1997). The low flow rates in the 10-100 nL/min range generated with microchips perfectly fit the sample flow requirement of ESI-MS (Feng and Smith 2000), despite the mismatch in physical dimensions between the bulky MS instrument and the small chip device. The microfluidic devices for ESI-MS can be classified into two categories with respect to purposes. The simpler type is just used for delivering and electrospraying samples into the mass spectrometer, i.e., infusion analysis. The second category is for coupling separation (CE, for example) to ESI-MS.

The main advantages of microfluidic devices over the conventional ESI interfaces arise from the way the ESI voltage is applied. Two typical layouts of microchips for ESI-MS are shown in Fig. 10. The electrospray voltage can be conveniently applied through a side channel which intersects the primary channel from which the analytes are sprayed (Fig. 10A) (Ramsey and Ramsey 1997; Xue et al. 1997). The easy estab-lishment of electrical connections through the channel network avoids the limitations associated with conventional ESI interfaces using liquid sheaths (Smith et al. 1993), liquid junctions (Severs and Smith 1997) and meal-coated capillary terminuses (Wahl et al. 1994). Figure 10B shows that a liquid junction structure is integrated in the chip with a minimized dead volume of 230 pL (Zhang et al. 2000). In infusion analysis the ESI voltage can be simply applied to the sample reservoir (Liu et al. 2000). Sample flow rates of 100-200 nL/min in the electrospray channel can be hydrodynamically introduced to the mass spectrometer using a syringe pump (Xue et al. 1997) or gas pressure at the channel inlet (Liu et al. 2000). Low flow rates of 18-30 nL/min, which are sufficient to maintain a stable electrospray, can be induced simply by an ESI voltage of 1.6 kV applied at the channel inlet (Lazar et al. 1999) .Larger EOF-driven flow rates (60-200 nL/min) can be generated with a controlled electric field strength (300-600 V/cm) by applying two voltages at the channel inlet and outlet, respectively (Lazar et al. 1999; Ramsey and Ramsey 1997).

It has been demonstrated that stable electrospray can be generated directly from the channel opening at the edge of a glass microchip without the assistance of any sheath liquid or sheath gas (Fig. 10A) (Ramsey and Ramsey 1997; Xue et al. 1997). The channel outlet can be formed simply by breaking the chip. It is interesting to note that one group found hydrophobic coatings around the channel opening helpful to achieve a stable electrospray (Xue et al. 1997) while the other did not (Ramsey and Ramsey 1997). Nevertheless, the liquid droplets (~10 nL) (Ramsey and Ramsey 1997) formed at the flat edge and the tendency of liquids to spread over the flat surface result in large dead volumes, which would preclude coupling to any on-chip separation. To remove the extra-column volume an integrated pneumatic nebulizer with two gas channels merging at the ESI channel opening has been developed (Zhang et al. 1999). The sample flow of 200 nL/min is sucked by the nebulizer with N_2 delivered at 0.3 L/min. Although CE-ESI-MS of peptide mixtures has been demonstrated using such a single piece of chip incorporating a pneumatic nebulizer, the separation efficiency achieved is inferior to that using an attached capillary as the emitter. In the devices with attached emitters, a fused-silica capillary (1-3 cm in length) is glued to the exit of the chip channel (Fig. 10B) (Lazar et al. 1999; Liu et al. 2000). In an impressive approach using a plastic chip of 16x10x0.9 cm with 96 microfabricated channels and 96 glued capillary emitters, 96 protein samples can be sequentially electrosprayed into the mass spectrometer in infusion mode without cross contamination (Liu et al. 2000). However, massive and tender labour must be involved in the gluing. It has recently been reported that the capillary emitter can be inserted into a guiding channel micromachined on the chip without gluing for easy replacement of either the capillary tip or the chip (Fig. 10B) (Wahl et al. 1994). In another approach a low dead volume connection procedure has been developed by drilling a flat-bottom hole instead of a tapered hole on the chip for receiving the capillary emitter (Bings et al. 1999).

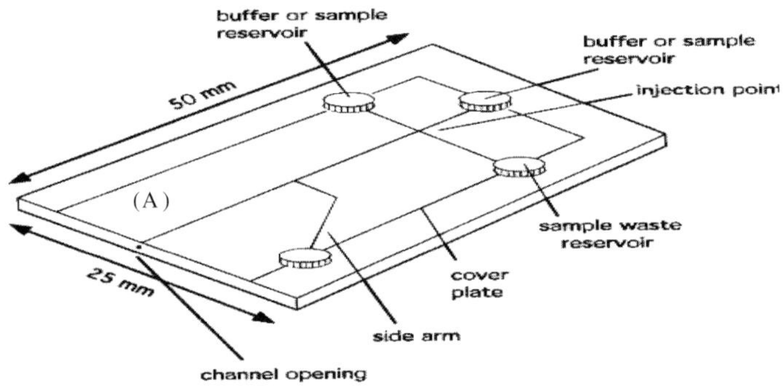

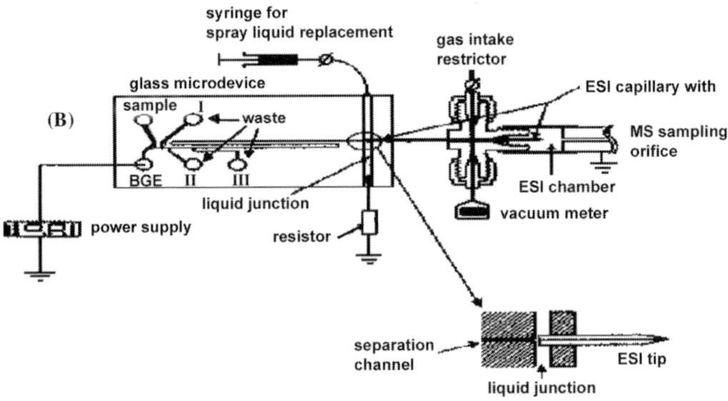

Fig. 10. Schematic diagrams of two ESI-MS chips. **(A)** A chip for direct electrospray from the channel opening on the chip edge. Voltages applied at the buffer reservoir and the side arm serve for electrokinetic pumping the sample and ESI voltage. (Reprinted from Xiang et al. 1999 with permission.) **(B)** A chip with an integrated liquid junction structure and a replaceable capillary emitter. The emitter tip is placed in a vacuum chamber at the orifice of the MS sucking the sample. (Reprinted from Woolley et al. 1997 with permission.)

To solve the connection problems associated with the attached capillaries an ideal solution would be to use integrated emitters. Needle-like structures protruding 1 mm outside the chip edge have recently been fabricated on a silicon wafer as a part of the chip (Fig. 11) (Licklider et al. 2000). In addition to the integrated emitters, sample and reagent wells with filters are also fabricated on chip. The chip has been tested in an infusion mode by applying gas pressures at the sample wells inducing sample flow rates of 35-77 nL/min without the use of any sheath liquid or gas. The ESI performance using the integrated chip is comparable to that using commercial pulled fused-silica capillary emitters. In another approach a polycarbonate chip with a micromachined emitter on it has been proved suitable for electrospray protein samples at flow rates of 50-200 nL/min with the assistance of sheath gas and liquid (Mao and Pawliszyn 1999). The fabrication of integrated emitters is just in its infancy and has yet to dominate in ESI-MS chips.

To date, most applications of ESI-MS chips focus on infusion analysis, including the impressive 96-sample system (Liu et al. 2000) and the dual microdialysis chip (Xiang et al. 2000). In the latter approach protein biomarkers for microorganism identification can be picked up by directly infusing crude cell lysates through the chip and on-chip electrospray of the dialyzed samples. Although MS has the ability to analyze several components in infusion modes, separation is generally required prior to MS analysis of real world samples when hundreds and thousands of compounds must be identified. However, only a few approaches to on-chip separation-electrospray have been reported, including CZE-ESI-MS (Zhang et al. 2000) and CIEF-ESI-MS (Zhang et al. 2000). Using a CZE-ESI-MS chip peptide mixtures are separated with a high plate number, 31 000, in the 11 cm separation channel (Zhang et al. 2000). Please note that this efficiency was obtained with a hydrodynamic flow of 100 nL/min sucked by the MS vacuum. Higher efficiency should be achievable without the hydrodynamic flow. In another approach, the plate number achieved in microchip CZE-ESI-MS of peptides is less than 3500 for a 5 cm separation channel because only one voltage is applied at the inlet buffer reservoir serving for both the separation and ESI (Li et al. 1999). In this case most of the voltage potential drop is between the emitter tip and the MS orifice resulting in a small electric field strength along the separation channel. In an approach of CIEF-ESI-MS on-chip focused protein bands are mobilized at a flow rate of 50 nL/min and sprayed using sheath liquid and sheath gas (Mao and Pawliszyn 1999).

It is important to note that several groups have been pursuing approaches using external liquid junction interfaces (Figeys and Aebersold 1998; Figeys et al. 1998), liquid sheath interfaces or metal-coated emitters (i.e., nano-electrospray emitters) (Li et al. 2000) to connect microfluidic devices. In this kind of approach the chips serve only for sample delivery (Figeys et al. 1998), mixing buffers (Figeys and Aebersold 1998) or part of the separation channel (Li et al. 2000). Although the liquid handling (sample dispense, injection, solvent delivery) can be facilitated using chips, the advantages offered by channel networks are not exploited for improved ESI efficiency. We classify this type of approach as sample handling rather than ESI-MS chips as no ESI is performed with the chips. This kind of approach inherits fully the limitations associated with the conventional ESI interfaces. For example, the metal-coated emitters suffer from the fact that air bubbles generated at the tip by electrochemical reactions often abruptly terminate electrospray, not to mention the limited life and tedious fabrication and handling. Therefore, even in conventional approaches without the use of chips metal-coated emitters should be avoided whenever they can (Feng and Smith 2000) .Liquid sheath and liquid junction interfaces are known to lead to reduced sensitivity (Severs and Smith 1997; Smith et al. 1993). In fact, these limitations associated with conventional ESI interfaces can be largely removed by use of microfluidic devices. On-chip microstructures equivalent to conventional liquid junction (Zhang et al. 2000) and ESI emitters (Licklider et al. 2000) have been realized.

Aiding liquids (Li et al. 1999) and gases (Zhang et al. 1999), when necessary, can be delivered through side channels. Another advantage is the applicability of on-chip coupling to separation and sample handling functions with minimized dead volumes. The nature of the microfluidic system makes it a potentially powerful tool for on-line sample pretreatment-separation-ESI.

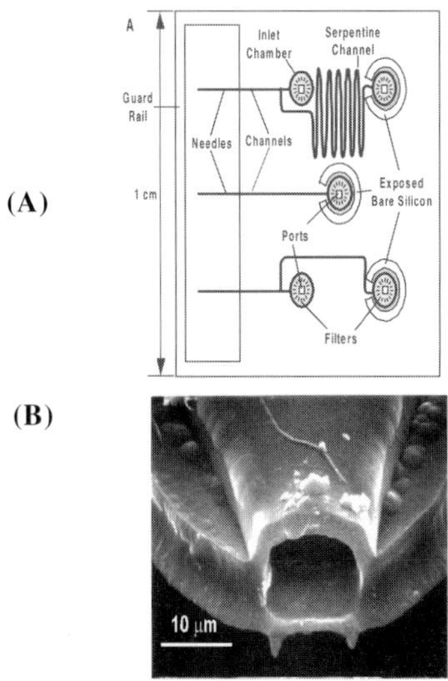

Fig. 11. (A) The layout of an ESI chip with integrated ESI emitters. **(B)** A scanning electron micrograph of the emitter tip. (Reprinted from [21] with permission.)

6. Conclusions and Outlook

At the current stage most of these investigations focus on implementing single analytical functions in a microfluidic device, making the bricks to build a mansion: a µTAS. The degree of functional integration is steadily increasing but just in its infancy. While academic society is seeking the limits of the microchips and trying everything in a chip format, the commercial sector focuses on the practical needs of the real world: the customers' needs. The ever broken limits in science and technology, which are always exciting but not necessarily beneficial to the real world, lead to increasing room for product development. The ultimate driving force of a new technique is the customers' choice. Every well developed technique possess at least one successful application area. A new technique has to beat the existing competitive one in overall performance against the costs of development. Here a question arises as to what benefits and new opportunities the chip technology can bring to our community that is already full of `advanced technologies'? The fundamental merits of microfluidic devices come from the small size and feasibility of integration. The nature of microfluidic devices has offered an exclusive opportunity to form a multiplexed and integrated system in one piece of chip: it is the ultimate driving force. The importance of microfluidic systems in respect to multiplexing and integration in revolutionizing analytical chemistry can be comparable to that of electronic chips in computer development.

On the routes towards the multiplexed and integrated systems choices have to be made. After a decade of exploring, some have and have not become obvious. (1) The voltage-driven separation techniques (CE) are obviously superior to their pressure-driven counterparts (HPLC) in chip formats, despite the conventional HPLC still playing a key role in pharmaceutical science and many other sciences. Not all conventional techniques share equal chances of success in miniaturized systems. (2) High pressure and mechanically moving parts, pumps and valves, should be avoided whenever they can. (3) Taking the nature of microchips into account can lead to extra improvement in performance when transforming a conventional method. A miniaturized system can work in quite a different way from its macro counterpart. For example, the microchannel network has offered exclusive advantages for performing ESI and coupling to on-chip separation. On the other hand, a miniaturized system resulting from simply scaling down or assembling the conventional systems on a piece of chip has few advantages over the parent techniques. Transformation of conventional techniques to chip formats is not always straightforward. (4) Incorporating real sample handling function into the microfluidic system is imperative. Sample handling is the most important interface between an operator and the system. For the system to be widely accepted, the interface must be user-friendly. (5) The sizes of the main channels at low micron levels should be small enough for most applications in the real world. However, sub-channels of low nm dimensions are needed for filtration and dialysis, which challenges presently available fabrication techniques. Fabrication of nanochannels with unduly high costs incurred will give way to in situ polymerization for forming pore structures for sample treatment.

Acknowledgements

We thank Zhala Sabir for reading this manuscript and for helpful comments.

References

Ahn CH, Puntambekar A, Lee SM, Cho HJ and Hong C-C (2000) Proc. µTAS 2000 Symposium In: Berg A van den, Olthuis W and Bergveld P (eds) Kluwer Academic Publishers, Dordrecht, pp 205
Anazawa T, Takahashi S and Kambara H (1996) Anal Chem 68:699
Arora A, Mello AJ de and Manz A (1997) Analytical Communications 34:393

Barker SLR, Tarlov MJ, Branham M, Xu J, MacCrehan W, Gaitan M and Locascio L E (2000) Proc. µTAS 2000 Symposium. In: Berg A van den, Olthuis W and Bergveld P (eds) Kluwer Academic Publishers, Dordrecht, pp 67

Berg A van den (2000) Proc. µTAS 2000 Symposium. In: Berg A van den, Olthuis W and Bergveld P (eds) Kluwer Academic Publishers, Dordrecht , pp vii

Bessoth FG, deMello AJ and Manz A (1999) Anal Commun 36: 213

Bings NH, Wang C, Skinner CD, Colyer CL, Thibault P and Harrison DJ (1999) Anal Chem 71: 3292

Braithwaite A and Smith FJ (1996) Chromatographic Methods 5th Ed. Blackie Academic & Professional, Glassgow, pp 33, 276, 357

Bruin GJM, Tock PPH, Kraak JC and Poppe H (1990) J Chromatogr 517: 557

Ceriotti L, Verpoorte E and Rooij NF de (2000) Proc. µTAS 2000 Symposium In: Berg A van den, Olthuis W and Bergveld P (eds) Kluwer Academic Publishers, Dordrecht, pp 225

Cheng J, Sheldon EL, Wu L, Uribe A, Gerrue LO, Heller M and O'Connell J (1998) Nat Biotech 16: 541

Cheng J, Waters LC, Fortina P, Hvichia G, Jacobson SC, Ramsey JM, Kricka LJ and Wilding P (1998) Anal Biochem 257: 101

Cohen CB, Chin-Dixon E, Jeong S and Nikiforov TT (1999) Anal Biochem 273: 89

Cole RB (ed) (1997) Electrospray Ionization Mass Spectrometry John Wiley & Sons, New York

Crabtree HJ, Kopp MU and Manz A (1999) Anal Chem 71: 2130

Culbertson CT, Jacobson SC and Ramsey JM (1998) Anal Chem 70:3781

Culbertson CT, Jacobson SC and Ramsey JM (2000) Proc µTAS 2000 Symposium. In: Berg A van den, Olthuis W and Bergveld P (eds) Kluwer Academic Publishers, Dordrecht, pp 221

Deshpande M, Greiner KB, West J, Gilbert JR, Bousse L and Minalla A (2000) Proc µTAS 2000 Symposium In: Berg A van den, Olthuis W and Bergveld P (eds) Kluwer Academic Publishers, Dordrecht, pp 339

Duffy DC, McDonald JC, Schueller OJA and Whitesides GM (1998) Anal Chem 70: 4974

Effenhauser CS, Bruin GJM, Paulus A and Ehrat M (1997) Anal Chem 69:3451

Effenhauser CS, Manz A and Widmer HM (1993) Anal Chem 65: 2637

Effenhauser CS, Paulus A, Manz A and Widmer HM (1994) Anal Chem 66: 2949

Ehrlich DJ and Matsudaira P (1999) TIBTECH 17: 315

Eijkel JCT, Mello A J de and Manz A (1999) Organic Mesoscopic Chemistry. In: Masuhara H and Schryver FC de (eds) Blackwell, Oxford, pp 185

Eijkel JCT, Stoeri H and Manz A (1999) Anal Chem 71: 2600

Erbacher C, Bessoth FG, Busch M, Verpoorte E and Manz A (1999) Mikrochim. Acta, 131: 19

Ericson C, Holm J, Ericson T and Hjertén S (2000) Anal Chem 72: 81

Feng B and Smith RD (2000) J Am Soc Mass Spectr 11:94

Figeys D and Aebersold R (1998) Anal Chem 70: 3721

Figeys D, Gygi SP, McKinnon G and Aebersold R (1998) Anal Chem 70: 3728

Fintschenko Y, Ngola SM, Shepodd TJ and Arnold DW (2000) Proc. µTAS 2000 Symposium In: Berg A van den, Olthuis W and Bergveld P (eds) Kluwer Academic Publishers, Dordrecht, pp 411

Frye-Mason G, Kottenstette R, Lewis P, Heller E, Manginell R, Adkins D, Dulleck G, Martinez D, Sasaki D, Mowry C, Matzke C and Anderson L (2000) Proc. µTAS 2000

Symposium. In: Berg A van den, Olthuis W and Bergveld P (eds) Kluwer Academic Publishers, Dordrecht, pp 229

Gao Q, Pang H-M and Yeung ES (1999) Electrophoresis 20: 1518

Gao Q and Yeung ES (2000) Anal Chem 72: 2499

Gelashvili DB, Tumanov AA, Bezrukov ME, Lisenkova NV, Barinova OV and Krest'yaninov PA (1999) J Anal Chem 54: 806

Gong X and Yeung ES (1999) Anal Chem 71: 4989

Harrison DJ, Fluri K, Seiler K, Fan Z, Effenhauser CS and Manz A (1993) Science, 261: 895

Harrison DJ, Manz A, Fan Z, Lüdi H and Widmer HM (1992) Anal Chem 64: 1926

Harvey D (2000) Modern Analytical Chemistry McGraw-Hill, Boston

Hashimoto M, Tsukagoshi K, Nakajima R, Kondo K and Arai A (2000) J Chromatogr A867: 271

He B, Ji J and Regnier FE (1999) J Chromatogr A853: 257

He B and Regnier F (1998) J Pharm Biomed Anal 17:925

He B, Tait N and Regnier F (1998) Anal Chem 70: 3790

He B, Tan L and Regnier F (1999) Anal Chem 71: 1464

Heeren F von, Verpoorte E, Manz A and Thormann W (1996a) J Microcolumn Separat 8: 373

Heeren F von, Verpoorte E, Manz A and Thormann W (1996b) Anal.Chem 68: 2044

Heiger DN (1992) High Performance Capillary Electrophoresis 2^{nd}Ed, Hewlett-Packard, Waldbronn

Hofmann O, Che D, Cruickshank KA and Müller UR (1999) Anal Chem 71: 678

Holm J, Gösch M, Blom H, Heino T, Thyberg P, Björk G and Rigler R (2000) Proc. µTAS 2000 Symposium. In: Berg A van den, Olthuis W and Bergveld P (eds) Kluwer Academic Publishers, Dordrecht, pp 273

Huang X, Coleman WF and Zare RNJ (1989) J Chromatogr 480: 95

Huang XC, Quesada MA and RA Mathies (1992) Anal Chem 64: 2149

Hutt LD, Glavin AP, Bada JL and RA Mathies (1999) Anal Chem 71: 4000

Ishizuka N, Minakuchi H, Nakanishi K, Soga N, Nagayama H, Hosoya K and Tanaka N (2000) Anal Chem 72: 1275

Jacobson SC, Ermakov SV and Ramsey JM (1999) Anal Chem 71: 3273

Jacobson SC, McKnight TE and Ramsey JM (1999) Anal Chem 71: 4455

Jakeway SC, Mello AJ de and Russell EL, Fresenius J (2000) Anal Chem 66: 525

Jiang G, Attiya S, Pcvirk G, Lee WE and Harrison DJ (2000) Biosensors & Bioelectronics 14:861

Jo B-H, Moorthy J and Beebe DJ (2000) Proc µTAS 2000 Symposium In: Berg A van den, Olthuis W and Bergveld P (eds) Kluwer Academic Publishers, Dordrecht, pp335

Kang SH, Gong X and Yeung ES (2000) Anal Chem 72: 3014

Khandurina J, Jacobson SC, Waters LC, Foote RS and Ramsey JM (1999) Anal Chem 71: 1815

Kopf-Sill AR (2000) Proc. µTAS 2000 Symposium In: Berg A van den, Olthuis W and Bergveld P (eds) Kluwer Academic Publishers, Dordrecht , pp 233

Kopp MU, Mello AJ de and Manz A (1998) Science 280: 1046

Krejci M, Pechan Z and Deyl Z (1975) Liquid Column Chromatograpgy In: Deyl Z, Macek K andJanak J (eds) Elsevier, Amsterdam, pp125

Krylov SN and Dovichi NJ (2000) Anal Chem 72: 111R

Kutter JP, Jacobson SC and Ramsey JM (1997) Anal Chem 69: 5165

Kutter JP, Jacobson SC, Mastubara N and Ramsey JM (1998) Anal Chem 70: 3291

Kutter JP (2000) Trends in Anal Chem 19: 352

Kwok YC and Manz A (2000) Electrophoresis, submitted

Kwok YC and Manz A, in preparation

Lagally ET, Simpson PC and Mathies RA (2000) Sensors & Actuators B63: 138

Lazar IM, Ramsey RS, Sundberg S and Ramsey JM (1999) Anal Chem 71: 3627

Lehmann U, Krusemark O, Müller J, Vogel A and Binz D (2000) Proc. µTAS 2000 Symposium In: A van den Berg, Olthuis W and Bergveld P (eds) Kluwer Academic Publishers, Dordrecht, pp 167

Lichtenberg J, Daridon A, Verpoorte E and Rooij NF de (2000) Proc. µTAS 2000 Symposium In: Berg A van den, Olthuis W and Bergveld P (eds) Kluwer Academic Publishers, Dordrecht, pp 307

Licklider L, Wang X-Q, Desai A, Tai Y-C and Lee TD (2000) Anal Chem 72: 367

Li J, Kelly JF, Chemushevich I, Harrison DJ and Thibault P (2000) Anal Chem 72: 599

Li J, Thibault P, Bings NH, Skinner CD, Wang C, Colyer C and Harrison JD (1999) Anal Chem 71: 3036

Li J, Wang C, Kelly JF, Harrison DJ and Thibault P (2000) Electrophoresis 21: 198

Linden H van der, Olthuis W and Bergveld P (2000) Proc µTAS 2000 Symposium In: Berg A van den, Olthuis W and Bergveld P (eds) Kluwer Academic Publishers, Dordrecht, pp 175

Liu H, Felten C, Xue Q, Zhang B, Jedrzejewski P, Karger BL and Foret F (2000) Anal. Chem 72: 3303

Liu RH, Yu Q, Bauer JM, Jo B-H, Moore JS and Beebe DJ (2000) Proc. µTAS 2000 Symposium In: Berg A van den, Olthuis W and Bergveld P (eds) Kluwer Academic Publishers, Dordrecht, pp 45

Liu S, Shi Y, Ja WW and Mathies RA (1999) Anal Chem 71: 566

Liu Y, Foote RS, Jacobson SC, Ramsey RS and Ramsey JM (2000) Proc. µTAS 2000 Symposium In: Berg A van den, Olthuis W and Bergveld P (eds) Kluwer Academic Publishers, Dordrecht, pp 295

Liu Y and Kuhr WG (1999) Anal Chem 71: 1668

MacNair JE, Patel KD and Jorgenson JW (1999) Anal Chem 71: 700

Ma L, Gong X and Yeung ES (2000) Anal Chem 72: 3383

Manz A, Grabe N and Widmer HM (1990) Sens Actuat B1: 244

Manz A, Harrison DJ, Verpoorte E and Widmer HM (1993) Advances in Chromatography In: Brown PR and Crushka E (eds) Volume 33 Marcel Dekker, New York, pp 1

Manz A, Harrison DJ, Verpoorte EMJ, Fettinger JC, Paulus A, Lüdi H and Widmer HM (1992) J Chromatogr 593:253

Manz A, Miyahara Y, Miura J, Watanabe Y, Miyagi H and Sato K (1990) Sensors & Actuators B1: 249

Mao Q and Pawliszyn J (1999) Analyst 124: 637

Marshell A and Hodgson J (1998) Nat Biotech 16:37

McEnery M, Tan A, Alderman J, Patterson J, O'Mathuna SC and Glennon JD (2000) Analyst 125:25

Mere L, Bennett T, Coassin P, England P, Hamman B, Rink T, Zimmerman S and Negulescu P (1999) DDT 4: 363

Mitchell MC, Spikmans V, Bessoth F, Manz A and Mello A de (2000) Proc. µTAS 2000 Symposium In: Berg A van den, Olthuis W and Bergveld P (eds) Kluwer Academic Publishers, Dordrecht, pp 463

Molho JI, Herr AE, Mosier BP, Santiago JG, Kenny TW, Brennen RA and Gordon GB (2000) Proc. μTAS 2000 Symposium In: Berg A van den, Olthuis W and Bergveld P (eds) Kluwer Academic Publishers, Dordrecht, pp 287

Mosher RA, Saville DA and Thormann W (1992) The Dynamics of Electrophoresis In: Radola BJ (ed) VCH, Weinheim, pp 11

Mueller O, Hahnenberger K, Dittmann M, Yee H, Dubrow R, Nagle R and Ilsley D (2000) Electrophoresis 21: 128

Mullis KB, Ferre F and Gibbs RA (1994) The Polymerase Chain Reaction, Birkhauser, Boston

Nishimoto T, Fujiyama Y, Abe H, Kanai M, Nakanishi H and Arai A (2000) Proc. μTAS 2000 Symposium In: Berg A van den, Olthuis W and Bergveld P (eds) Kluwer Academic Publishers, Dordrecht, pp 395

Ocvirk G, Munroe M, Tang T, Oleschuk R, Westra K and Harrison DJ (2000) Electrophoresis 21: 107

Ocvirk G, Verpoorte E, Manz A, Grasserbauer M and Widmer HM (1995) Anal Methods Instrum 2: 74

Oleschuk RD and Harrison DJ (2000) Trends in Anal Chem 19: 379

Oleschuk RD, Shultz-Lockyear LL, Ning Y and Harrison DJ (2000) Anal Chem 72: 585

Quirino JP, Terabe S and Bocek P (2000) Anal Chem 72: 1934

Paegel BM, Hutt LD, Simpson PC and Mathies RA (2000) Anal Chem 72: 3030

Prazen BJ, Bruckner CA, Synovec RE and Kowalski BR (1999) Anal Chem 71:1093

Preisler J and Yeung ES (1996) Anal Chem 68: 2885

Prest JE, Baldock SJ, Bektas N, Fielden PR and Brown BJT (1999) J Chromatogr A 836: 59

Ramseier A, Heeren F von and Thormann W (1998) Electrophoresis 19:2967

Ramsey RS and Ramsey JM (1997) Anal Chem 69: 1174

Rodriguez-Diaz R, Wehr T and Zhu M (1997) Electrophoresis 18: 2134

Salas-Solano O, Schmalzing D, Koutny L, Buonocore S, Adourian A, P Matsudaira and D Ehrlich (2000) Anal Chem 72: 3129

Salimi-Moosavi H, Jiang YT, Lester L, McKinnon G and Harrison DJ (2000) Electrophoresis 21: 1291

Sanders GHW and Manz A (2000) Trends in Anal Chem 19: 364

Schmalzing D, Adourian A, Koutny L, Ziaugra L, Matsudaira P and Ehrlich D (1998) Anal Chem 70: 2303

Schmalzing D, Koutny L, Chisholm D, Adourian A, Matsudaira P and Ehrlich D (1999) Anal Biochem 270: 148

Schrum DP, Culbertson CT, Jacobson SC and Ramsey JM (1999) Anal Chem 71: 173

Severs JC and Smith RD (1997) Anal Chem 69: 2154

Shi Y, Simpson PC, Scherer JR, Wexler D, Skibola C, Smith MT and Mathies RA (1999) Anal Chem 71: 5354

Simpson PC, Roach D, Woolley AT, Thorsen T, Johnston R, Sensabaugh GF and Mathies RA (1998) Proc. Natl Acad Sci 95: 2256

Sirichai S and Mello A J de (2000) Analyst 125: 133

Smith N and Evans MB (1999) J Chromatogr A32: 41

Smith RD, Wahl JH, Goodlett DR and Hofstadler SA (1993) Anal Chem A 65: 574

Sullivan MG, Utomo H, Fagan PJ and Ward MD (1999) Anal Chem 71: 4369[102]

Swanson P, Gelbart R, Atlas E, Yang L, Grogan T, Bulter WF, Ackley DE and Sheldon E (2000) Sens Actuat B64: 22

Tantra R and Manz A (2000) Anal Chem 72: 2875

Tashiro K, Sekiguchi T, Shoji S, Funatsu T, Masumoto W and Sat Ho (2000) Proc. µTAS 2000 Symposium In: Berg A van den, Olthuis W and Bergveld P (eds) Kluwer Academic Publishers, Dordrecht, pp 209

Terrett NK (1998) Combinatorial Chemsitry, Oxford University Press, Oxford

Terry SC, Jerman JH, Angell JB (1979) IEEE Trans Electron Devices ED-26: 1880

Thormann W, Zhang C-X, Caslavska J, Gebauer P and Mosher RA (1998) Anal Chem 70: 549

Tian H, Jaquins-Gerstl A, Munro N, Trucco M, Brody LC and Lander JPs (2000) Genomics 63: 25

Todd MD, Lin XD, Stankowski LF, Desai M and Wolfgang GHI (1999) J Biomed Screening 4: 259

Ueno K and Yeung ES (1994) Anal Chem 66: 1424

Unger MA, Chou H-P, Thorsen T, Scherer A and Quake SR (2000) Science 288: 113

Venter JG, Adams MD, Sutton GG, Kerlavage AR, Smith HO and Hunkapiller M (1998) Science 280: 1540

Verpoorte EMJ, Schoot BH van der, Jeanneret S, Manz A, Widmer HM and de Rooij NF (1994) J Micromech Microeng 4: 246

Vo-Dinh T, Alarie JP, Isola N, Landis D, Wintenberg AL and Ericson MN (1999) Anal Chem 71: 358

Wahl JH, Gale DC and Smith RD (1994) J Chromatogr A659: 217

Wallenborg SR and Bailey CG (2000) Anal Chem 72: 1872

Wang J, Chatrathi MP and Tian B (2000) Analytical Chimica Acta 416: 9

Waterston R and Sulston J E (1998) Science 282 53

Wen J, Lin Y, Xiang F, Matson DW, Udseth HR and Smith RD (2000) Electrophoresis 21: 191

Wilding P, Kricka, LJ, Cheng J, Hvichia G, Shoffner MA and Fortina P (1998) Anal Biochem 257: 95

Woias P, Hauser K and Yacoub-George E (2000) Proc. µTAS 2000 Symposium In: Berg A van den, Olthuis W and Bergveld P (eds) Kluwer Academic Publishers, Dordrecht, pp 277

Woolley AT, Sensabaugh GF and Mathies RA (1997) Anal Chem 69:2181

Xiang F, Anderson GA, Veenstra TD, Lipton MS and Smith RD (2000) Anal Chem 72: 2475

Xiang F, Lin Y, Wen J, Matson DW and Smith RD (1999) Anal Chem 71: 1485

Xue Q, Foret F, Dunayevskiy YM, Zavracky PM, McGruer NE and Karger BL, (1997) Anal Chem 69: 426

Xu Y, Bessoth FG, Eijkel JCT and Manz A (2000) Analyst 125: 677

Zeng L and Kassel DB (1998) Anal Chem 70: 4380

Zhang B, Foret F and Karger BL (2000) Anal Chem 72: 1015

Zhang B, Liu H, Karger BL and Foret F (1999) Anal Chem 71: 3258

Zhang C-X, Sun Z-P and Ling D-K (1993) J Chromatogr A655: 309

Zhang C-X and Thormann W (1998) Anal Chem 70:540

Zhang C-X and Thormann W (1996) Anal Chem 68: 2523

Zhang C-X, Xiang F, Paša-Tolić L, Anderson GA, Veenstra TD and Smith RD, (2000) Anal Chem 72: 1462